Palgrave Studies in the History of Science and Technology

Series Editors

James Rodger Fleming
Colby College
Waterville, Maine, USA

Roger D. Launius
Smithsonian Institution
National Air and Space Museum, USA

Designed to bridge the gap between the history of science and the history of technology, this series publishes the best new work by promising and accomplished authors in both areas. In particular, it offers historical perspectives on issues of current and ongoing concern, provides international and global perspectives on scientific issues, and encourages productive communication between historians and practicing scientists.

More information about this series at
http://www.springer.com/series/14581

Thomas Heinze • Richard Münch
Editors

Innovation in Science and Organizational Renewal

Historical and Sociological Perspectives

Editors
Thomas Heinze
Interdisciplinary Center for Science
and Technology Studies (IZWT)
University of Wuppertal
Wuppertal, Germany

Richard Münch
University of Bamberg
Bamberg, Germany

Palgrave Studies in the History of Science and Technology
ISBN 978-1-137-59419-8   ISBN 978-1-137-59420-4   (eBook)
DOI 10.1057/978-1-137-59420-4

Library of Congress Control Number: 2016944788

Cover illustration: © StudioM1/Thinkstock

Printed on acid-free paper

This Palgrave Macmillan imprint is published by Springer Nature
The registered company is Nature America Inc. New York

CONTENTS

CONTRIBUTORS

David Baneke is Assistant Professor of History of Science at Utrecht University. His main research interests are history of astronomy and the role of science in modern society. He recently published *De Ontdekkers van de Hemel* (2015), a history of Dutch astronomy in the twentieth century.

Irwin Feller is Professor Emeritus of Economics at the Pennsylvania State University and former Visiting Scientist at the American Association for the Advancement of Science. Among his published works are studies of the diffusion of innovation, science policy, higher education, interdisciplinary research, and program evaluation.

Edward J. Hackett is vice provost for research and a professor in the Heller School of Social Policy and Management at Brandeis Univesity. Prof. Hackett studies the organization and dynamics of collaborations in various forms and places, and has written about peer review, science policy, and the scientific career. He is also editor of Science, Technology & Human Values.

Olof Hallonsten is a senior researcher in the Department of Business Administration at Lund University and also affiliated with the KTH Royal Institute of Technology in Stockholm, Sweden. He specializes in the study of the governance and organization of large research organizations ("Big Science") and is author of several important articles on a variety of aspects on this subject.

Thomas Heinze is Professor of Organizational Sociology and member of the Interdisciplinary Center for Science and Technology Studies (IZWT) at Wuppertal University, Germany. His research interests include scientific creativity, research breakthroughs, governance of research organizations, research evaluation, theories of institutional change, and organizational renewal.

Arlette Jappe is a senior researcher at the Institute of Sociology and the Interdisciplinary Center for Science and Technology Studies (IZWT) at Wuppertal University, Germany. Her research interests are in sociology of science, research evaluation, and environmental sociology.

Roger D. Launius is Associate Director for Collections and Curatorial Affairs at the Smithsonian Institution's National Air and Space Museum, Washington, DC. He is the editor of *Exploring the Solar System: The History and Science of Planetary Exploration* (2013, Palgrave) and several other works.

Cyrus C. M. Mody is Professor of History of Science, Technology, and Innovation at Maastricht University. He specializes in the recent history of the physical and engineering sciences in USA, with special emphasis on nanotechnology, commercialization of academic research, and university–industry–government collaboration. He is the author of *Instrumental Community: Probe Microscopy and the Path to Nanotechnology* (2011).

Richard Münch is Professor Emeritus of Sociology at the University of Bamberg, Germany, and Senior Professor of Social Theory and Comparative Macrosociology at Zeppelin University Friedrichshafen, Germany. He is a member of the Berlin Brandenburg Academy of Sciences. Among his recent publications is *Academic Capitalism. Universities in the Global Struggle for Excellence* (2014).

John N. Parker is a fellow at Barrett, The Honors College at Arizona State University, and associate research scientist at the National Center for Ecological Analysis and Synthesis at the University of California at Santa Barbara. His research focuses on the sociology of science, creativity, emotions, and small groups and has been published in venues such as *American Sociological Review, Social Studies of Science*, and *Bioscience*.

Roger H. Stuewer is Professor Emeritus of the History of Science and Technology at the University of Minnesota in Minneapolis. He has published extensively on the history of quantum and nuclear physics. He was awarded the American Physical Society and American Institute of Physics Abraham Pais Prize for History of Physics in 2013, the Distinguished Alumni Award of the Department of Physics of the University of Wisconsin in Madison in 2014, and an American Association of Physics Teachers Homer L. Dodge Distinguished Service Citation in 2016.

CHAPTER 1

Editors' Introduction: Institutional Conditions for Progress and Renewal in Science

Thomas Heinze and Richard Münch

1.1 PROGRESS AND RENEWAL IN SCIENCE

In the history, philosophy, and sociology of science, there is a consensus that the primary goal of scientific research is the continuous renewal of knowledge and technology. In this context, renewal refers not only to the generation of new ideas, theories, methods, and instruments or to the discovery of previously unknown phenomena but also to the diffusion of innovative scientific developments, and the institutionalization of such advances in existing scientific communities and ultimately as new academic fields. Accepting the premise that the renewal of knowledge and

T. Heinze (✉)
University of Wuppertal, Wuppertal, Germany

R. Münch
University of Bamberg, Bamberg, Germany

1
T. Heinze, R. Münch (eds.), *Innovation in Science and Organizational Renewal*, DOI 10.1057/978-1-137-59420-4_1

technology is the objective of scientific research, we can then ask what are institutional conditions for successful renewal. This edited volume contributes to the debate about renewal in science by addressing two interrelated questions. First, this volume explores the capability of research organizations to generate original and transformative intellectual contributions, such as new theories, methods, instrumentation, and empirical discoveries. Second, this volume addresses the capability of national research systems and research organizations to absorb new intellectual developments and to institutionalize new fields of research. Through detailed historical and comparative case studies, this volume presents new and thought-provoking evidence that improves our conceptual knowledge and empirical understanding about how new research fields are formed, how research organizations adapt to changes both in the sciences and in their societal environment, and how research sponsors strike the balance between support for new research areas and continuity for established lines of disciplinary research.

Investigating the complex connections between scientific innovation and institutional change requires a long-term perspective. Therefore, the volume assembles scholars in science history, as well as in sociology of science and research policy. Yet, the distinctive contribution of this volume is that while being firmly based in science history, it strives for broader and more general sociological and policy propositions regarding renewal in science. Through the juxtaposition between science history and the sociology of science and research policy, we attempt to narrow the gap between detailed microhistories of particular entities or episodes and over-generalized sociological propositions on institutional change in science.

In this introductory chapter, we argue that renewal within the organizations that conduct scientific research, as well as within their environment, is contingent upon at least three institutional conditions: *(1) investments in exploration, (2) facilitation of meso-level competition, and (3) organizing interdisciplinary research.* What follows below is a discussion of these three institutional conditions, how each chapter in this edited volume contributes to their analysis, and finally, extended abstracts of all chapters.

1.2   Investments in Exploration

Generally speaking, scientists face two opposing expectations. First, they are expected to seek fundamentally new knowledge and to move beyond established doctrine. Second, they are expected to develop and maintain an inventory of disciplinary knowledge that can be passed on from generation to generation. These two expectations are conflicting, and they operate as antipodal values under various labels: innovation versus tradition, originality versus relevance, dissent versus conformity, rebellion versus discipline, exploration versus exploitation, search versus production, experimentation versus implementation, or risk taking versus refinement.

Michael Polanyi argues that the tension between these two opposing expectations pervades the entire institutional structure of scientific research: "This internal tension is essential in guiding and motivating scientific work. The professional standards of science must impose a framework of discipline and at the same time encourage rebellion against it. They must demand that ... an investigation should largely conform to the currently predominant beliefs about the nature of things, while allowing in order to be original it may to some extent go against these."[1]

That there is a fundamental tension between seeking new and refining existing knowledge implies that depending on historical circumstances and institutional context there may either be a delicate balance between the two, or one pole will dominate the other. Polanyi argues that the institutional structure of science—in general—tends to be biased toward the refinement of existing knowledge. Taking peer review as an example, he claims that publications are primarily evaluated in terms of their plausibility and scientific value, and thus with respect to their contribution toward an inventory of disciplinary knowledge. Publications have to be plausible and valuable extensions of existing knowledge for them to be accepted by the scientific community. In contrast, publications of sufficient plausibility and scientific value may vary considerably with respect to their originality, that is, the degree of surprise which they arouse among scientists. Hence, not every publication, no matter how plausible and valuable it may be, is novel and original.

In a similar vein, Richard Whitley argues that despite the strong institutional commitment to the exploration of fundamentally new knowledge in modern science, "the extent of originality and novelty in research goals and procedures is restricted by the need to convince specialist colleagues of the significance of one's work in reputational work organizations. ... The degree of innovation is thus diminished and constrained by the necessity of showing how new contributions fit in with, and are relevant to, existing knowledge."[2] Hence, Whitley asserts that the scientific elite holds the innovators in check. Novel ideas and artifacts are accepted only if they can be connected to previous knowledge and thus prove their scientific relevance. The view that the institutional structure of science gives considerably more weight to the plausibility of contributions and their connectability to previous research than to originality and surprise has garnered empirical support in recent years. Many commentators argue that during the past three decades, the funding of public research organizations has increasingly shifted toward external, peer-reviewed sponsorship despite that such funding tends to favor mainstream and risk-averse projects.[3] Thus, the proliferation of peer review in funding decisions most likely has deepened existing knowledge paths at the expense of finding fundamentally new ones.

In his essay on exploration versus exploitation in organizational learning, James March warns that "systems that engage in exploitation to the exclusion of exploration are likely to find themselves trapped in suboptimal stable equilibria." He concludes that "maintaining an appropriate balance between exploration and exploitation is a primary factor in system survival and prosperity."[4] In this respect, it is interesting that several private and public research sponsors, among them the Volkswagen Foundation, the Wellcome Trust, the MacDonnell Foundation, the Howard Hughes Medical Institute, and the European Research Council, set up funding programs dedicated to the support of unconventional research that has the potential for groundbreaking results.[5] Many of these programs are intended to counterbalance the dominant exploitation-mode inherent in research council funding. Yet, typically they command small budgets, operate under heightened evaluation requirements, and rely a fortiori on traditional peer review.[6]

The two observations in the literature that the institutional structure of science tends to be biased toward the refinement of existing knowledge, and that research funding in recent decades has strengthened established knowledge paths have led sociologists of science and organizational scholars alike to reconsider institutional conditions in support of explorative and path-breaking research. The common theme in these contributions is that the forces of exploration need to be strengthened to balance the two conflicting orientations in the institutional structure of science. This plea for *investments in exploration* is articulated either from a comparative historical perspective,[7] from an organizational sociology perspective,[8] from an individual's research strategy view,[9] or from a research policy viewpoint.[10]

The present volume contributes to this renewed discussion by asking (1) how and why *investments in exploration* have occurred historically and (2) more generally, how the two opposing orientations of innovation and tradition are balanced in different institutional settings. In contrast to the current emphasis on funding structure, this volume puts emphasis on new organizational forms and internal organizational change. Several chapters in this volume present evidence that *investments in exploration* are made by building entirely new forms of research organizations, such as the university-based microfabrication user facility (Mody), the National Center for Ecological Analysis and Synthesis (Hackett and Parker), and the space research laboratories and consortia that built the two satellites ANS and IRAS (Baneke); or new forms of conferences, such as the Solvay Conferences or the Seven Pines Symposia (Stuewer). These new organizations or conferences are examples of an ongoing process of renewal in the institutional arrangements of science that have considerable effects on intellectual opportunities and innovations. In addition, several chapters in this volume present cases of adaptation and internal change of existing research organizations, including the Deutsches Elektronensynchrotron (DESY) and the Stanford Linear Accelerator Center (SLAC) (Hallonsten and Heinze), or the Goddard Space Flight Center at the National Aeronautics and Space Administration, NASA (Launius). As shown by these chapters, internal organizational changes oftentimes occur gradually, particularly in institutional environments in which entrance of new forms of research organizations is either difficult or impossible, or in cases where existing research capacities can serve as platform for building new ones. Therefore, both founding new organizational forms and supporting gradual internal adaptations of existing research organizations are two equally important investments in exploration.

1.3   FACILITATION OF MESO-LEVEL COMPETITION

In addition to the tension between exploration and exploitation, *competition* pervades the entire institutional structure of scientific research. A classical view on competition in science is Karl Popper's falsificationist account on how theories are used to explain phenomena and to make forecasts.[11] If a theory fails to explain or forecast a phenomenon, this may constitute an anomaly that has no immediate impact on the theory. However, frequent occurrence of such anomalies weakens a theory's foundation. As soon as a new theory is available that is more successful at explaining and predicting observed events, the old theory should be abandoned. Thus, the driving force of scientific progress is *competition between theories* for better explanations.

In his discussion of Popper's approach to scientific progress, Imre Lakatos points out that theories must first be constructed and then initially protected against criticism, since otherwise they would be abandoned before fully blooming.[12] The establishment of research programs serves precisely this goal. Such programs pursue a specific knowledge goal in a given field of research, using a particular set of theoretical basic assumptions and methods. According to Lakatos, the differentiation of research programs into a protected core of basic assumptions and a peripheral area of special hypotheses produces a balance between stability and change that serves progress in knowledge better than ubiquitous and aggressive criticism. It can also be considered beneficial for scientific progress when the protagonists of a research program do their utmost to protect their program against possible criticism, and leave it to their competitors to launch criticism and offer alternatives. Hence, in Lakatos' view, competition between theories is less important than *competition between research programs.*

According to classical sociology of science, the competition between either theories or research programs is socially embedded in scientific fields where scientists compete for reputation and intellectual control.[13] In this regard, Whitley points out that "scientific fields are a particular kind of work organization which structure and control the production of intellectual novelty through competition for reputations from national and international audiences for contributions to collective goals."[14] However, scientists are not just seeking personal acclaim from colleagues for their scientific achievements, "they also seek to direct others' research along particular lines and ensure that their interests, problems, and standards are accepted by colleagues in their own research."[15]

In addition to the argument that individual scientists seek reputation and intellectual control, sociology of science discusses how *nation states compete for global leadership in science and technology.* For example, Joseph Ben-David demonstrates that ever since the emergence of the modern sciences in the seventeenth century in Renaissance Italy, competition for global scientific and technological leadership has been a driving force in science.[16] More recent history and sociology of science studies corroborate this view in that such international competition has influenced the emergence of new science and technology fields, particularly during the Cold War, including fields such as earth sciences, space science, ocean-ography, seismology, and biotechnology.[17]

This brief outline illustrates that competition in science is a multilevel phenomenon, including cognitive aspects, such as competition between theories or research programs, and social and historical aspects, such as individuals competing with colleagues for scientific reputation and intel-lectual control or nation states competing for global scientific, military, and technological leadership. However, the *meso-level of research organiza-tions* has been largely neglected in scholarly discourse on competition as an institutional condition for scientific progress and renewal. Although some studies have shown that both the distribution of scientific productivity and the number of major scientific achievements are highly skewed among universities and non-university research laboratories,[18] we know relatively little about the institutional conditions that increase the scientific competi-tiveness of universities and other public and private research laboratories, neither do we know much about capabilities of research systems to flexibly adapt their organizational infrastructure to heightened global scientific and technological competition.

Therefore, the present volume aims at contributing to a better under-standing of *meso-level competition* in science by asking (1) which factors are conducive to research organizations' capabilities to seize upon new scientific opportunities, and thus successfully compete in emerging fields of science and technology, and (2) how new research capacities are built up to strengthen national competitiveness in response to global scientific and technological pressures.

Several chapters in this volume present evidence in this regard: in a comparison between public universities in Germany and the USA, it is shown that the capability of universities to support new fields of research critically depends on both their funding and scientific staff structures (Jappe and Heinze); furthermore, it is demonstrated that inter-university competition was a major driver in the proliferation of the microfabrication user facility in the USA, and that the leading contenders in this competi-tion were universities that could demonstrate a long-term track record of partnership with industry (Mody); yet another chapter argues that the small community of Dutch astronomers forged an alliance between policy makers and two major Dutch companies, Philips and Fokker, to build very expensive scientific instruments (satellites), and thereby consider-ably improved their global scientific and technological competitiveness (Baneke).

1.4   ORGANIZING INTERDISCIPLINARY RESEARCH

In addition to investments in exploration and the facilitation of meso-level competition, the *emergence of new disciplines and specialties* is often regarded as emblematic for progress and renewal in science. In his late writings, Thomas Kuhn argues that similar to speciation of new biological organisms, new disciplines emerge when scientists increasingly rely on a new lexicon that excludes non-specialists from scientific communication. Therefore, breakdowns in communication between scientists are "crucial symptoms of the speciation-like process through which new disciplines emerge, each with its own lexicon, and each with its own area of knowledge."[19] Most importantly, Kuhn argues that "very likely it is the specialization consequent on lexical diversity that permits the sciences, viewed collectively, to solve the puzzles posed by a wider range of natural phenomena than a lexically homogeneous science could achieve."[20] Hence, the increasing specialization of lexicons reduces communication between different research areas, but at the same time, it increases the diversity of scientific approaches, and thus our knowledge to understand the (physical) world. It is by the division of specialized scientific communication that knowledge grows: "the limited range of possible partners for fruitful intercourse is the essential precondition for what is known as progress in both biological development and the development of knowledge."[21]

Kuhn's strong emphasis on incommensurability between disciplinary lexicons as a prerequisite for scientific progress and renewal can be contrasted with the concept of intellectual "trading zones"[22] which instead focuses on "interdisciplinary partnership in which two or more perspectives are combined and a new, shared language develops."[23] Quite in general, studies in interdisciplinarity, multidisciplinarity, and transdisciplinarity agree that disconnected branches of scientific research can be effectively linked.[24]

Evidence from science history and the sociology of science suggests that both private and public sponsorship, and the establishment of new types of research institutes both inside and outside universities, played an important role in effectively *organizing interdisciplinary research*. For example, Ben-David argues that interdisciplinary research centers in universities in the USA, established across discipline-based departments, were more successful scientifically than discipline-based institutes that prevailed in Germany.[25] Both David A. Hounshell and John Kenly Smith and John W. Servos show that the emergence of physical chemistry as a new field of research was supported by the fruitful application of physics tools and techniques to chemistry, and sponsored by large chemical corporations both within their own laboratories and through grants to major research universities.[26] In addition, Robert E. Kohler describes the Rockefeller Foundation's dedication to funding scientists who applied the tools and techniques of physics and chemistry toward the advancement of knowledge of biological processes, and how this played an important role in building research capacity in molecular biology.[27] More recent studies, including J. Rogers Hollingsworth, Jerald Hage, and Jonathon Mote, suggest that research laboratories, which were internally structured into groups rather than discipline-based departments, were highly effective in establishing productive work relationships between scientists from various specialties and fields.[28]

In light of the discussion above, the contribution of this edited volume is threefold. First, it assembles contributions that provide considerable support for the argument that effective communication across disciplinary boundaries is facilitated by new types or forms of research organizations: the university-based microfabrication user facility (Mody), the National Center for Ecological Analysis and Synthesis, and the Resilience Alliance (Hackett and Parker) are recent examples that add further substance to the existing literature.

Second, several chapters show research organizations have considerable adaptive capabilities when research across disciplinary boundaries is required: NASA established, alongside its main mission, a multidisciplinary earth system science program (Launius); DESY and SLAC gradually replaced particle physics by the study of materials by X–rays as the main purpose of accelerators, and established units for multidisciplinary photon science inside their formal organizational structure (Hallonsten and Heinze); the two companies Philips and Fokker, together with several Dutch university institutes, were engaged in research consortia to which scientific and engineering staff from various disciplines was recruited for conducting space-related research and development (Baneke).

Third, several chapters argue that interdisciplinarity is anchored not only in centers or institutes but also in scientific careers: NASA encouraged many individuals to migrate from planetary to earth science, helping to create earth science as a cohesive entity (Launius); abundant research opportunities in emerging scientific fields that were adjacent to where scientists had worked before, provided the opportunity structure to effectively link different methods and competences (Jappe and Heinze);

and academic astronomers were in a good position to move into space research, provided they were able to attract people with technological and managerial competence into their research groups (Baneke).

Last but not least, the final chapter in this volume shows how interdisciplinary research has been initiated and shaped by national science policy, and that recent shifts toward funding interdisciplinary research at the expense of mainstream disciplinary research, and increasing requirements for accountability and evidence of performance on the part of those receiving public-sector support have produced tighter funding conditions for academic researchers, even as total science-agency budgets have increased (Feller).

1.5   CONTRIBUTIONS TO THE EDITED BOOK

Cyrus Mody's *Fabricating an Organizational Field for Research: US Academic Microfabrication Facilities in the 1970s and 1980s* (Chap. 2) examines the emergence and diffusion of the university-based microfabrication user facility in the USA. This new organizational form arose in the 1970s to foster greater interaction among stakeholders in industry, academia, and government, thereby facilitating new and innovative research in materials science. Mody describes the mechanisms by which this new organizational form was replicated and spread, and how it coevolved with shifts in industrial structure, including the decline of basic research in semiconductor companies, as well as shifts in federal science policy, primarily the decline of defense-related R&D. This new type of facility diffused widely in the USA today constitutes an entire organizational field of its own.

Edward Hackett and John Parker's *From Salomon's House to Synthesis Centers* (Chap. 3) analyzes synthesis centers as an innovative form of scientific organization that promotes the integration of scientific diversity and its engagement with real-world problems. Placed in historical perspective, such centers are examples of an ongoing process of renewal in the organizational and institutional arrangements of science, and they have consequences for the character and effects of scientific knowledge. Hackett and Parker describe how intellectual and institutional innovations emerge and are entwined within such centers, then draw upon ideas from science studies, small group dynamics, and the creativity and interdisciplinarity literatures to identify the patterns and processes of social interaction responsible for the centers' performance.

Roger Stuewer's *The Seventh Solvay Conference: Nuclear Physics, Intellectual Migration, and Institutional Influence* (Chap. 4) demonstrates how new types of conferences promote mutual learning of scientists from different national and institutional contexts. The chapter shows how the seventh Solvay Conference in 1933 lay at the crossroads in the history of experimental and theoretical nuclear physics when new experimental techniques and instruments were being developed and new theoretical ideas and concepts were being generated, all of which were diffused to physicists in many countries of the world. Stuewer shows the influence that the Solvay Conferences exerted as a model for future conferences in physics and in the history and philosophy of physics, particularly the Seven Pines Symposia.

Olof Hallonsten and Thomas Heinze's *"Preservation of the Laboratory is not a Mission." Gradual Organizational Renewal in National Laboratories in Germany and the United States* (Chap. 5) examines the gradual but transformative changes inside two national laboratories in the USA (SLAC) and Germany (DESY) from single-mission particle physics laboratories in the early 1960s to multipurpose research centers for photon science in the 2000s. The authors describe how the field of synchrotron radiation research increasingly challenged, and ultimately succeeded, particle physics as the established discipline in these laboratories. Their focus is on the processes that led to intra-organizational change, including conversion of large technical infrastructures, gradual replacement of particle physics by the study of materials by X–rays as the main purpose of accelerators, and layering of new organizational units for photon science. By investigating the complexity of institutional change at the micro-level of two laboratories, the chapter contributes important conceptual tools for a more detailed understanding of organizational adaptation and renewal.

Arlette Jappe and Thomas Heinze's *Institutional Context and Growth of New Research Fields. Comparison between Universities in Germany and the United States* (Chap. 6) shows that differences in funding and staff structure of state universities in Germany and the USA affect the capabilities of their research groups and departments to rapidly seize upon research breakthroughs. Using the Scanning Tunneling Microscope, STM (Nobel Prize in Physics, 1986) and the discovery of Buckminster Fullerenes, BUF (Nobel Prize in Chemistry, 1996) as empirical examples, they demonstrate that universities whose budgets grew and had a high number of professors among their scientific staff were among the early adopters of STM and BUF, and thus highly competitive in the newly emerging research fields.

In contrast, universities whose budgets stagnated and had a low share of professors among their scientific staff were mostly among those who engaged in follow-up research relatively late.

David Baneke's *Organizing Space: Dutch Space Science between Astronomy, Industry and the Government* (Chap. 7) shows that whenever new technological or scientific fields emerged after the Second World War, scientists, government officials, and industrial companies in the Netherlands feared being left behind. Especially in strategically important fields such as nuclear physics, radio astronomy, and computing, these three groups collaborated intensively to keep up with international developments; and Philips as a major company played an important role in these collaborations. Using space science as an example, Baneke demonstrates how the small community of Dutch astronomers, with the help of Philips and Fokker, managed to build two of the most expensive scientific instruments ever built in the Netherlands: the two satellites ANS and IRAS. The new research capacities that were created both in Dutch universities and in Philip's and Fokker's laboratories considerably improved the scientific and technological competitiveness of the Netherlands.

Roger Launius's *"We will learn more about the Earth by leaving it than by remaining on it." NASA and the Forming of an Earth Science Discipline in the 1960s* (Chap. 8) argues that despite recent criticism that NASA in the 1960s failed to recognize and make a part of its core mission "earthly environmentalism," this chapter responds by discussing the manner in which NASA in a subtle but transformative way encouraged the collaboration of scientists from many different disciplines focused on Earth to transcend disciplinary boundaries using space technology to treat Earth as an integrated system. Indeed, from limited cooperative efforts in the 1960s overseen by NASA, emerged the broadly interdisciplinary efforts to understand the interactions of Earth in the last quarter century. While such efforts never dominated the agency and were resisted in some quarters, the seeds of the earth system science discipline were planted during this era. Launius shows that NASA encouraged many individuals to migrate from planetary to earth science, helping to create earth science as a cohesive entity.

Irwin Feller's *Interdisciplinary Research and Transformative Research as Facets of National Science Policy* (Chap. 9) argues that the total resources required to satisfy the claims for continued support of established academic disciplines on the one hand, and for underwriting the reconfigurations of these disciplines into new research fields on the other hand, push up against and invariably exceed whatever level of total resources are provided by the collectivity of sponsors. Therefore, the strong emphasis in the USA's national science policy on interdisciplinary research, and more recently on transformative research, is emblematic for the ongoing debate about how important public-science funding should be, and what levels and forms of funding are most appropriate. Feller argues that recent shifts toward funding interdisciplinary research at the expense of mainstream disciplinary research, and increasing requirements for accountability and evidence of performance might lead to adverse conditions for academic researchers in disciplinary settings, even as total science-agency budgets continue to increase in absolute terms.

NOTES

1. Michael Polanyi, *Knowing And Being*. With an Introduction by Marjorie Grene (Chicago: Chicago University Press, 1969), 55.
2. Richard Whitley. *The Intellectual and Social Organization of the Sciences*. Second Edition (Oxford: Oxford University Press, 2000), 28.
3. Frank A. Zoller, Eric Zimmerling and Roman Boutellier, "Assessing the Impact of the Funding Environment on Researchers' Risk Aversion: the Use of Citation Statistics," *Higher Education* 68 (2014): 333–345; Christine Musselin, "How Peer Review empowers the Academic Profession and University Managers: Changes in Relationships between the State, Universities and the Professoriate," *Research Policy* 42 (2013): 1165–1173; Thed N. van Leeuwen and Henk F. Moed, "Funding Decisions, Peer Review, and Scientific Excellence in Physical Sciences, Chemistry, and Geosciences," *Research Evaluation* 21 (2012): 189–198; Lutz Bornmann and Hans-Dieter Daniel, "The Manuscript Reviewing Process: Empirical Research on Review Requests, Review Sequences, and Decision Rules in Peer Review," *Library and Information Science Research* 32 (2010): 5–12; Grit Laudel, "The Art of Getting Funded: How Scientists Adapt to Their Funding Conditions," *Science and Public Policy* 33 (2006): 489–504; Liv Langfeldt, "The Decision-Making Constraints and Processes of Grant Peer Review, and Their Effects on the Review Outcome," *Social Studies of Science* 31 (2001): 820–841; Paul Bourke and Linda Butler, "The Efficacy of Different Modes of Funding Research: Perspectives from Australian Data on the Biological Sciences," *Research Policy* 28 (1999): 489–499; Daryl E. Chubin and Edward J. Hackett, *Peerless Science: Peer Review and U.S. Science Policy* (Albany, N.Y.: State University of New York Press, 1990); Fiona Wood and Simon Wessely, "Peer Review of Grant Applications," in *Peer Review in Health Science:* 2nd Edition, ed. Fiona Godlee and Tom Jefferson (British Medical Association Publications, 2003): 14–44.

4. James G. March, "Exploration and Exploitation in Organizational Learning," *Organization Science* 2 (1991): 71.

5. Tertu Luukkonen, "The European Research Council and the European Research Funding Landscape," *Science and Public Policy* 41 (2014): 29–43; Thomas Heinze, "How to Sponsor Ground-Breaking Research: A Comparison of Funding Schemes," *Science and Public Policy* 35 (2008): 302–318; Patrick J. Prendergast, Sheena H. Brown and J.R. Britton, "Research Programmes that Promote Novel, Ambitious, Unconventional and High-Risk Research: an Analysis," *Industry and Higher Education* 22 (2008): 215–221; Jonathan Grant and Liz Allen, "Evaluating High Risk Research: An Assessment of the Wellcome Trust's Sir Henry Wellcome Commemorative Awards for Innovative Research," *Research Evaluation* 8 (1999): 201–204.

6. Heinze, "Sponsor Ground-Breaking Research"; Grant and Allen, "Evaluating High Risk Research."

7. Richard Münch, *Academic Capitalism: Universities in the Global Struggle for Excellence* (London/New York: Routledge, 2014); Jerald Hage and Jonathon Mote, "Transformational Organizations and Institutional Change: the Case of the Institut Pasteur and French Science," *Socio-Economic Review* 6 (2008): 313–336; J. Rogers Hollingsworth, "A Path-Dependent Perspective on Institutional and Organizational Factors Shaping Major Scientific Discoveries," in *Innovation, Science, and Institutional Change*, ed. Jerald Hage and Marius Meeus (Oxford: Oxford University Press, 2006), 423–442; Robert K. Merton and Elinor G. Barber, *The Travels and Adventures of Serendipity: A Study in Sociological Semantics and the Sociology of Science* (Princeton, N.J.: Princeton University Press, 2004).

8. Jan Youtie et al., "Career-based Influences on Scientific Recognition in the United States and Europe: Longitudinal Evidence from Curriculum Vitae Data," *Research Policy* 42 (2013): 1341–1355; Thomas Heinze et al., "Organizational and Institutional Influences on Creativity in Scientific Research," *Research Policy* 38 (2009): 610–623; Richard Whitley, "Changing Governance of the Public Sciences," in *The Changing Governance of the Sciences*, ed. Richard Whitley and Jochen Gläser (Dordrecht: Springer, 2007), 3–27.

9. Jacob G. Foster, Andrey Rzhetsky and James A. Evans, "Tradition and Innovation in Scientists' Research Strategies," *American Sociological Review* 80 (2015): 875–908.

10. Zoller, Zimmerling and Boutellier, "Assessing the Impact"; Laudel, "The Art of Getting Funded"; Sven Hemlin, Carl M. Allwood and Ben R. Martin, *Creative Knowledge Environments: The Influences on Creativity in Research and Innovation* (Cheltenham: Edward Elgar, 2004); Dietmar

Braun, "The role of funding agencies in the cognitive development of science," *Research Policy* 27 (1998): 807–821.

11. Karl R. Popper, *Conjectures and Refutations: The Growth of Scientific Knowledge* (London: Routledge, 2002[1963]).

12. Imre Lakatos, "Criticism and the Methodology of Scientific Research Programmes," *Proceedings of the Aristotelian Society: New Series* 69 (1968/1969): 149–186; Imre Lakatos, "The Methodology of Scientific Research Programmes" in *Philosophical Papers: Volume 1*, ed. John Worrall and Gregory Currie (Cambridge: Cambridge University Press, 1978).

13. Whitley, *The Intellectual and Social Organization*; Robert K. Merton, Harriet Zuckerman, "Institutionalized Patterns of Evaluation in Science," in *The Sociology of Science: Theoretical and Empirical Investigations*, ed. Robert K. Merton and Harriet Zuckerman (Glencoe: Free Press, 1973), 460–496; Jerry Gaston, *Originality and Competition in Science* (Chicago: Chicago University Press, 1973).

14. Whitley, *The Intellectual and Social Organization*, 80.

15. Ibid., 97.

16. Joseph Ben-David, *The Scientist's Role in Society* (Chicago: University of Chicago Press, 1971).

17. Ronald E. Doel and Kristine C. Harper, "Prometheus Unleashed. Science as a Diplomatic Weapon in the Lyndon B. Johnson Administration," *Osiris* 21 (2006): 66–85; Sheila Jasanoff, "Biotechnology and Empire. The Global Power of Seeds and Science," *Osiris* 21(2006): 273–292; Jacob Darwin Hamblin, *Oceanographers and the Cold War: Disciples of Marine Science* (Seattle: University of Washington Press, 2005); Ronald E. Doel, "Constituting the Postwar Earth Sciences: The Military's Influence on the Environmental Sciences in the USA after 1945," *Social Studies of Science* 33(2003): 635–666.

18. Benjamin F. Jones, Stefan Wuchty and Brian Uzzi, "Multi-University Research Teams: Shifting Impact, Geography, and Stratification in Science," *Science* 322 (2008): 1259–1262; Hage and Mote, "Transformational Organizations"; Hollingsworth, "A Path-Dependent Perspective"; J. Rogers Hollingsworth, "Institutionalizing Excellence in Biomedical Research: The Case of The Rockefeller University," in *Creating a Tradition*, ed. D.H. Stapleton (New York: Rockefeller University Press, 2004), 17–63.

19. Thomas Kuhn, "The Road Since Structure," in *The Road Since Structure: Philosophical Essays, 1970–1993*, ed. James Conant and John Haugeland (Chicago & London: Chicago University Press, 2000), 100–101.

20. Kuhn, "The Road Since Structure," 99.

21. Ibid.

22. Harry M. Collins, Robert Evans and Mike Gorman, "Trading Zones and Interactional Expertise," *Studies in History and Philosophy of Science* 38 (2007): 657–666; Peter Galison, *Image and Logic: A material Culture of Microphysics* (Chicago: Chicago University Press, 1997).

23. Collins, Evans and Gorman, "Trading Zones", 657.

24. Robert Crease, "Physical Sciences," in *The Oxford Handbook of Interdisciplinarity*, ed. Robert Frodeman, Julie Thomson Klein, and Carl Mitcham (Oxford/New York: Oxford University Press, 2010), 79–102; Julie Thompson Klein, "A Taxonomy of Interdisciplinarity" in *The Oxford Handbook of Interdisciplinarity*, ed. Robert Frodeman, Julie Thomson Klein, and Carl Mitcham (Oxford/New York: Oxford University Press, 2010), 15–30; Alan Porter, Ismael Rafols, "Is Science Becoming More Interdisciplinary? Measuring and Mapping Six Research Fields over Time," *Scientometrics* 81 (2009): 719–745; Wesley Shrum, Joel Genuth and Ivan Chompalov, *Structures of Scientific Collaboration* (Cambridge: MIT Press, 2007).

25. Joseph Ben-David, *The Scientist's Role*.

26. David A. Hounshell and John Kenly Smith, *Science and Corporate Strategy: Du Pont R & D 1902–1980* (Cambridge: Cambridge University Press, 1988); John W. Servos, *Physical Chemistry from Ostwald to Pauling: The Making of a Science in America* (Princeton NJ: Princeton University Press, 1990).

27. Robert E. Kohler, *Partners in Science: Foundations and Natural Scientists 1900–1945* (Chicago/London: University of Chicago Press, 1991).

28. Hollingsworth, "Institutionalizing Excellence"; Hollingsworth, "A Path-Dependent Perspective"; Jerald Hage, "Organizations and Innovation: Contributions from Organizational Sociology and Administrative Science," in *Innovation and Institutions. A Multidisciplinary Review of the Study of Innovation Systems*, ed. Steven Casper and Frans van Waarden (Cheltenham: Edward Elgar, 2005), 71–112; Hage and Mote, "Transformational Organizations."

Acknowledgments Several chapters in this volume were presented as papers at the "International Conference on Intellectual and Institutional Innovation in Science," held at the Berlin-Brandenburg Academy of Sciences and Humanities, September 13–15, 2012. We are very grateful to the conference committee members for reviewing papers; these include (in alphabetical order): Mats Benner, Dietmar Braun, Susan Cozzens, Ronald Doel, James Evans, Jacob Hamblin, Stefan Kuhlmann, Jacques Mairesse, Patrick McCray, Ben Martin, Christine Musselin, Dominique Pestre, Philip Shapira, and Richard Whitley. The conference was sponsored by the German Federal

Ministry for Education and Research (Bundesministerium für Bildung und Forschung, BMBF) as part of grant 01UZ1001: special thanks to Dietrich Nelle (Head of Section 42, BMBF) and Monika Wächter (PT–DLR). Regarding the edited volume, we are very grateful for helpful comments and suggestions from two anonymous reviewers, and also many thanks to Steffi Heinecke and David Pithan for editing the book manuscript.

REFERENCES

Ben-David, Joseph. 1971. *The scientist's role in society.* Chicago: University of Chicago Press.

Bornmann, Lutz, and Hans-Dieter Daniel. 2010. The manuscript reviewing process: Empirical research on review requests, review sequences, and decision rules in peer review. *Library and Information Science Research* 32: 5–12.

Bourke, Paul, and Linda Butler. 1999. The efficacy of different modes of funding research: Perspectives from Australian data on the biological sciences. *Research Policy* 28: 489–499.

Braun, Dietmar. 1998. The role of funding agencies in the cognitive development of science. *Research Policy* 27: 807–821.

Chubin, Daryl E., and Edward J. Hackett. 1990. *Peerless science: Peer review and U.S. science policy.* Albany: State University of New York Press.

Collins, Harry M., Robert Evans, and Mike Gorman. 2007. Trading zones and interactional expertise. *Studies in History and Philosophy of Science* 38: 657–666.

Crease, Robert. 2010. Physical sciences. In *The Oxford handbook of interdisciplinarity*, ed. Robert Frodeman, Julie Thompson Klein, and Carl Mitcham, 79–102. Oxford/New York: Oxford University Press.

Doel, Ronald E. 2003. Constituting the postwar earth sciences: The military's influence on the environmental sciences in the USA after 1945. *Social Studies of Science* 33: 635–666.

Doel, Ronald E., and Kristine C. Harper. 2006. Prometheus unleashed. Science as a diplomatic weapon in the Lyndon B. Johnson administration. *Osiris* 21: 66–85.

Foster, Jacob G, Andrey Rzhetsky, and James A. Evans. 2015. Tradition and Innovation in Scientists' Research Strategies. *American Sociological Review* 80: 875–908.

Galison, Peter. 1997. *Image and logic: A material culture of microphysics.* Chicago: Chicago University Press.

Gaston, Jerry. 1973. *Originality and competition in science.* Chicago: Chicago University Press.

Grant, Jonathan, and Liz Allen. 1999. Evaluating high risk research: An assessment of the Wellcome Trust's Sir Henry Wellcome Commemorative Awards for Innovative Research. *Research Evaluation* 8: 201–204.

Hage, Jerald. 2005. Organizations and innovation: Contributions from organizational sociology and administrative science. In *Innovation and institutions: A multidisciplinary review of the study of innovation systems*, ed. Steven Casper and Frans van Waarden, 71–112. Cheltenham: Edward Elgar.

Hage, Jerald, and Jonathon Mote. 2008. Transformational organizations and institutional change: The case of the Institut Pasteur and French science. *Socio-Economic Review* 6: 313–336.

Hamblin, Jacob Darwin. 2005. *Oceanographers and the Cold War: Disciples of marine science.* Seattle: University of Washington Press.

Heinze, Thomas. 2008. How to sponsor ground-breaking research: A comparison of funding schemes. *Science and Public Policy* 35: 302–318.

Heinze, Thomas, Philip Shapira, Juan D. Rogers, and Jacqueline M. Senker. 2009. Organizational and institutional influences on creativity in scientific research. *Research Policy* 38: 610–623.

Hemlin, Sven, Carl M. Allwood, and Ben R. Martin. 2004. *Creative knowledge environments: The influences on creativity in research and innovation.* Cheltenham: Edward Elgar.

Hollingsworth, J. Rogers. 2004. Institutionalizing excellence in biomedical research: The case of The Rockefeller University. In *Creating a tradition*, ed. D.H. Stapleton, 17–63. New York: Rockefeller University Press.

Hollingsworth, J. Rogers. 2006. A path-dependent perspective on institutional and organizational factors shaping major scientific discoveries. In *Innovation, science, and institutional change*, ed. Jerald Hage and Marius Meeus, 423–442. Oxford: Oxford University Press.

Hounshell, David A., and John Kenly Smith. 1988. *Science and corporate strategy: Du Pont R & D 1902–1980.* Cambridge: Cambridge University Press.

Jasanoff, Sheila. 2006. Biotechnology and empire. The global power of seeds and science. *Osiris* 21: 273–292.

Jones, Benjamin F., Stefan Wuchty, and Brian Uzzi. 2008. Multi-university research teams: Shifting impact, geography, and stratification in science. *Science* 322: 1259–1262.

Klein, Julie Thompson. 2010. A taxonomy of interdisciplinarity. In *The Oxford handbook of interdisciplinarity*, ed. Robert Frodeman, Julie Thompson Klein, and Carl Mitcham, 15–30. Oxford/New York: Oxford University Press.

Kohler, Robert E. 1991. *Partners in science: Foundations and natural scientists 1900–1945.* Chicago/London: University of Chicago Press.

Kuhn, Thomas. 2000. The road since structure. In *The road since structure: Philosophical essays, 1970–1993*, ed. James Conant and John Haugeland, 90–104. Chicago/London: Chicago University Press.

Lakatos, Imre. 1968/1969. Criticism and the methodology of scientific research programmes. *Proceedings of the Aristotelian Society: New Series* 69: 149–186.

Lakatos, Imre. 1978. The methodology of scientific research programmes. In *Philosophical papers: Volume 1*, ed. John Worrall and Gregory Currie. Cambridge: Cambridge University Press.

Langfeldt, Liv. 2001. The decision-making constraints and processes of grant peer review, and their effects on the review outcome. *Social Studies of Science* 31: 820–841.

Laudel, Grit. 2006. The art of getting funded: How scientists adapt to their funding conditions. *Science and Public Policy* 33: 489–504.

Luukkonen, Tertu. 2014. The European Research Council and the European research funding landscape. *Science and Public Policy* 41: 29–43.

March, James G. 1991. Exploration and exploitation in organizational learning. *Organization Science* 2: 71–87.

Merton, Robert K., and Elinor G. Barber. 2004. *The travels and adventures of serendipity: A study in sociological semantics and the sociology of science*. Princeton: Princeton University Press.

Merton, Robert K., and Harriet Zuckerman. 1973. Institutionalized patterns of evaluation in science. In *The sociology of science: Theoretical and empirical investigations*, ed. Robert K. Merton and Harriet Zuckerman, 460–496. Glencoe: Free Press.

Münch, Richard. 2014. *Academic capitalism: Universities in the global struggle for excellence*. London/New York: Routledge.

Musselin, Christine. 2013. How peer review empowers the academic profession and university managers: Changes in relationships between the state, universities and the professoriate. *Research Policy* 42: 1165–1173.

Polanyi, Michael. 1969. *Knowing and being*. With an introduction by Marjorie Grene. Chicago: Chicago University Press.

Popper, Karl R. 2002[1963]. *Conjectures and refutations: The growth of scientific knowledge*. London: Routledge.

Porter, Alan, and Ismael Rafols. 2009. Is science becoming more interdisciplinary? Measuring and mapping six research fields over time. *Scientometrics* 81: 719–745.

Prendergast, Patrick J., Sheena H. Brown, and J.R. Britton. 2008. Research programmes that promote novel, ambitious, unconventional and high-risk research: An analysis. *Industry and Higher Education* 22: 215–221.

Servos, John W. 1990. *Physical chemistry from Ostwald to Pauling: The making of a science in America*. Princeton: Princeton University Press.

Shrum, Wesley, Joel Genuth, and Ivan Chompalov. 2007. *Structures of scientific collaboration*. Cambridge: MIT Press.

van Leeuwen, Thed N., and Henk F. Moed. 2012. Funding decisions, peer review, and scientific excellence in physical sciences, chemistry, and geosciences. *Research Evaluation* 21: 189–198.

Whitley, Richard. 2000. *The intellectual and social organization of the sciences*, 2nd ed. Oxford: Oxford University Press.

Whitley, Richard. 2007. Changing governance of the public sciences. In *The changing governance of the sciences*, ed. Richard Whitley and Jochen Gläser, 3–27. Dordrecht: Springer.

Youtie, Jan, Juan Rogers, Thomas Heinze, Philip Shapira, and Li Tang. 2013. Career-based influences on scientific recognition in the United States and Europe: Longitudinal evidence from curriculum vitae data. *Research Policy* 42: 1341–1355.

Zoller, Frank A., Eric Zimmerling, and Roman Boutellier. 2014. Assessing the impact of the funding environment on researchers' risk aversion: The use of citation statistics. *Higher Education* 68: 333–345.

CHAPTER 2

Fabricating an Organizational Field for Research: US Academic Microfabrication Facilities in the 1970s and 1980s

Cyrus C.M. Mody

2.1 INTRODUCTION

The research area today known as "nanofabrication" or "nanostructure fabrication" takes a very broad purview. As the description of the 2012 Gordon Research Conference on Nanostructure Fabrication put it, the field's jurisdiction includes:

> novel fabrication methods and the limits of lithography; single atom and molecule manipulation and devices; nano-electrical, –mechanical, and -optoelectronic devices and phenomena; fabrication involving nanostructured and atomic-scale materials, including nanowires, nanotubes, and graphene; biological and biomolecular assembly and fabrication at the nanoscale; and the physics and applications of such devices and structures.[1]

C.C.M. Mody (✉)
Maastricht University, Maastricht, Netherlands

© The Editor(s) (if applicable) and The Author(s) 2016
T. Heinze, R. Münch (eds.), *Innovation in Science and Organizational Renewal*, DOI 10.1057/978-1-137-59420-4_2

As that description implies, practitioners of nanofabrication hail from many disciplines and commercial sectors, and use a variety of techniques to characterize and manipulate many different nanoscale phenomena. That diversity was not always the case. The origins of this conference, and the field as a whole, date to the late 1960s and early 1970s. At that time, nanofabrication was "microfabrication," since its techniques could only make structures with dimensions around one micron (millionth of a meter), whereas today's techniques can make structures with nano-meter dimensions (billionths of a meter). In the micro era, practitio-ners oriented almost entirely to the microelectronics industry, utilized only a few techniques, and were trained primarily in electrical engineer-ing or applied physics. Microfabrication diversified in both techniques and disciplines as it moved into the nanoscale in the 1990s. Both trans-formations were enabled by post-Cold War reshaping of organizations and institutions—journals, research units, conferences series, funding streams—that supported microfabrication since the late 1960s.

This chapter traces the emergence and reshaping of a distinctive new organizational form associated with microfabrication—the university-based interdisciplinary microfabrication user facility. From the beginning of a microfabrication community, user facilities have shared equipment and technical advice with a large user base seeking to construct leading-edge microelectronic circuits and other nanoscale experimental devices. These facilities also played an important role in shifts in American science and science policy, since they have been used as models for university–industry–government partnerships since the late 1970s.

The history of academic microfabrication user facilities offers a window onto transitions in American (and global) science in the late twentieth century: less corporate basic research; more reliance on industrial research consortia and university–industry partnerships; more pressure on faculty to patent research, found start-up companies, and "translate" findings into civil society; less federal money for academic laboratory buildings. These transitions have been given suggestive labels such as "triple helix," "Mode 1/Mode 2," "post-academic science," "post-modern science," and "neo-liberal science," but their exact nature is still contentious.[2]

This chapter makes three contributions to understanding post-1980 science, especially in the USA. First, I call attention to the microelec-tronics industry's role in stimulating late and post-Cold War changes in American science. In most accounts of these changes, the biotech and pharmaceutical industries receive the lion's share of the blame/credit, but as *Science* put it in 1982,

> While attention has been focused on the expanding links between academic biologists and the corporate world, a second revolution in university-industry relationships has been taking place in a different field. Electronics companies, faced with growing competition from Japan and fearing a shortage of well-trained Ph.D.s, are pouring unprecedented amounts of cash into university electrical engineering and computer science departments.[3]

The microelectronics industry's influence was not a one-time event precipitated by the threat of competition from Japanese firms, however. In microfabrication research, continuing changes in the structure of the microelectronics industry were echoed by changes in what "microfabrication" meant, who was doing it, with what tools, and for what purposes. Thus, the second contribution of this chapter is to show how reorganization of an industry can lead to reorganization of a research field.

The final contribution is to offer a case study of how the new values, practices, and tools of post-1980 American science spread from campus to campus via the emergence of an "organizational field"—by which I mean "those organizations that, in the aggregate, constitute a recognized area of institutional life ... that produce similar services or products."[4] An organizational field of academic microfabrication user facilities became a recognized area of institutional life in the late 1970s by providing equipment and advice to researchers making experimental microelectronic (and other) devices.

The formation of organizational fields in science is a largely neglected topic among historians of the postwar era. For earlier periods, we know how new organizational forms such as universities and national institutes diffused.[5] We have excellent depictions of individual, exemplary postwar research units.[6] However, little has been written about how exemplary units *became* exemplary—how their practices and values spread to similar organizations. This is a woeful gap, considering the importance of networks of research centers as a tool of postwar (especially post-Sputnik) science policy.[7]

Microfabrication facilities were a much-copied instance of an organizational field in science. Many of these facilities spun off from earlier organizational fields centered on other areas of research, particularly materials science. A few (initially 5, later 14) microfabrication facilities eventually formalized as a federally funded network of centers. Even the facilities not formally in this network looked to other sites for insight into how to administer and operate such an organization. Thus, the emergence of the organizational field of microfabrication facilities aided in what the editors of this volume call "investments in exploration" via new organizational forms that bring about innovations in techniques (as methods and equipment propagated across the field) as well as in administrative procedures and even communal norms (such as the increasing value placed on interdisciplinary collaboration).

2.2   THE LITHOGRAPHY WARS

The first academic facilities with microfabrication equipment shared among multiple users were apparently founded at Berkeley in the early 1960s, and Stanford and the University of Arizona somewhat later.[8] There was also an earlier Massachusetts Institute of Technology (MIT) group led by Dudley Buck (affiliated with Project Whirlwind) that operated much like later facilities. Certainly, today's MIT nanofabrication facility celebrates Buck as a notable ancestor.[9]

These first facilities were founded during a dramatic transition in microelectronics technology and manufacturing. Electronic circuits improve with miniaturization—they get faster, cheaper, and more durable and energy-efficient. Even in the vacuum tube era, there were intense efforts to miniaturize circuits for military applications and hearing aids.[10] However, microfabrication, in the sense the term later acquired, only became possible with the invention of the integrated circuit in 1957. At that point, manufacturers, government researchers and grant and procurement officers, and academic physicists and electrical engineers realized vast improvements in circuit performance and cost could be achieved by improving the means of miniaturizing circuit patterns.

Over time, many lithographic techniques were proposed. By the late 1960s, though, a consensus formed that the microelectronics industry would rely primarily on techniques in which a particle beam shone through a template (a "mask") onto a wafer coated with a "resist," altering the resist to make it more (or less) susceptible to acid in places where the beam passed through the mask. Acid would then be applied to transfer the pattern on the mask onto the wafer. To make complex structures, the process would be repeated, sometimes with tens of masks.

But what kind of particle beam to use? Photon beams, ion beams, and electron beams all competed in "lithography wars" waged in conferences, laboratories, boardrooms, and journals.[11] Optical photon lithography led from the start, with electron beam lithography at its heels. By the mid-1970s, Bell Labs was using e-beam lithography to make masks, and many observers assumed "direct-write" e-beam would soon replace optical lithography for writing commercial chip patterns.[12] Further behind were ion beam and extreme ultraviolet photon beam lithographies.[13]

The late 1960s and early 1970s saw rapid progress in all these techniques. That progress gave impetus to—and was sustained by—the formation of institutions where these techniques' proponents could share knowledge. An annual "Three Beams" meeting took shape in the late 1960s, though it took several years to acquire a formal steering committee and begin publishing proceedings. In 1976, the Gordon Conference series on the Chemistry and Physics of Microstructure Fabrication began, overlapping the Three Beams meeting in leadership and attendance.[14] Similar conferences also sprouted in Europe and Japan.

These institutions arose because of the complex relationship among the three beams. Each beam had advocates who believed it would be the primary technique for making commercial integrated circuits. But each beam also relied on ancillary technologies that it shared with other beams: resists, equipment for cleaning wafers and growing crystals, steppers to align masks, and so on. At a few universities, uncertainty about which technique would triumph combined with knowledge spillover among different techniques gave rise to rudimentary microfabrication "facilities." Wherever multiple faculty members on one campus were working on different aspects of microfabrication, an incentive arose to share expensive equipment that the different microfabrication techniques had in common. Here, for instance, is one of the founding members of the University of Texas' Microelectronics Research Center (MRC) describing how things worked even before the MRC was founded in 1984:

> The photolithography and the thermal equipment are two easy things to identify as equipment that didn't make a whole lot of sense to be owned and solely used by one group. They're painful enough to keep and maintain and the duty cycle's low. So, those are naturally shared: photolithography, simple metallization, etch, and thermal furnaces. Those are four things that are just sort of naturals for co-owned or co-used equipment.[15]

Initially, pooling was a radical departure from academic norms of individual achievement. Here, for instance, is a description of the philosophy behind the Berkeley facility:

> One of the principles that [the founders] agreed upon was going to be one lab and they would all share as equal partners and if others wanted to join, they could be accepted as equal partners too—which was definitely in contrast to the pattern of most experimental academic programs, where it was Professor X's lab, right, and only X and his students, and everything was under his control.... [T]he reason a lot of people oppose [sharing], is that when you don't have enough of the right lab discipline, one person can screw up everything. Everything can be put out of order and damaged.[16]

Despite the dangers of pooled equipment falling prey to the tragedy of the commons, and of individuals' having to cede autonomy, the expense and diversity of microfabrication equipment were conducive to sharing.

2.3   NATIONAL RESEARCH AND RESOURCE FACILITY

The sense that research equipment was becoming so expensive that academic researchers were priced out of many fields—not just microfabrication—was keenly felt in science policy circles in the 1970s.[17] The National Science Foundation (NSF), in particular, responded with the concept of federally funded shared user facilities. In astronomy, for instance, the NSF established four "National Research Centers" to redress an imbalance between an elite corps of astronomers with access to a few private, first-class observatories, and the larger mass of practitioners with access to public, second-class equipment.[18] Similarly, in 1973, NSF provided funds for the Stanford Synchrotron Radiation Project to serve as a "national user facility" for materials science, microelectronics, and molecular biology.[19] Later in the decade, Berkeley's Museum of Vertebrate Zoology received a major Operational Support grant from the NSF's Biological Research Resources program to serve as a repository available to users from around the world.[20] In chemistry, the NSF received authorization in 1978 to fund a series of Regional Instrumentation Facilities, 14 of which were built by 1982.[21]

The NSF was at the time undergoing dramatic change. Prior to 1970, it was a small funder of individual basic researchers in the physical and life sciences. In the 1970s, it began to put more emphasis on engineering and social sciences and on interdisciplinary, applied research, and to channel more funding through centers rather than individuals. This was also when the Mansfield Amendment (barring the military from funding basic research) forced the Pentagon to transfer oversight and funding of several interdisciplinary academic research centers to the NSF, such as the dozen or so Materials Research Laboratories (MRLs) and the Francis Bitter National Magnet Laboratory at MIT.[22]

It was in this context that, in 1974, the NSF's 16 Engineering Division program officers were instructed to put forward proposals for a "marquee" project that would solidify the division's growing influence within the foundation.[23] One, Jay Harris, proposed that the NSF fund a university-based shared equipment facility for microfabrication, based on the model of the Bitter Magnet Lab.[24] As one of Harris' superiors, Charles Polk, head of the Engineering Division in 1976–77, put it:

> We have talked about a national center or several regional laboratories *where that major, expensive equipment would be available....* [T]he large initial investment and the continuing support which are required could be justified only in terms of benefits to many research workers and to many different institutions. As a consequence, a national or regional laboratory, supported by NSF, would have to make very good provisions for guest workers and would have to engage permanent personnel which would help visitors with physical implementation of their ideas.[25]

Harris' own experiences as a faculty member motivated his proposal:

> I used to visit various industrial laboratories to try to get some help in making small optical structures. I got my best reception at the Hughes research labs in Malibu, from a guy named Ed Wolf, who was working with electron beams, but Ed didn't really have time to devote to supporting academics trying to work over their heads.[26]

Wolf later concurred that by 1975:

> a very noticeable gap opened between university research on the one hand and the accomplishments of industrial laboratories on the other—a gap due mainly to the expensive equipment and the interdisciplinary nature of microstructure science and engineering that universities found difficult, if not impossible, to support.[27]

Harris' proposal was well timed because the institutionalization and technological progress of the microfabrication research community were already drawing the NSF's attention. But he was also lucky that his proposal coincided with the 1975 announcement by the Japanese Ministry of International Trade and Industry (MITI) of a crash program to aid Japanese semiconductor firms.[28] The resulting panic among American science policymakers would last another 15 years. At the outset of that panic, Harris' proposal allowed the NSF to be out in front of addressing declining US competitiveness in microelectronics.

Harris' proposal therefore doubly benefited from what the editors of this volume term "renewal in science." Harris' original goal was to contribute to a long-term, evolutionary renewal already underway in electrical engineering, applied physics, and microelectronics manufacturing—an evolutionary renewal made possible by developments in techniques for fabricating integrated circuits and by the emergence of institutions for supporting and propagating innovations in those techniques. After the 1975 MITI announcement, though, Harris' proposal was folded into what the editors of this volume call "facilitation of meso-level competition" via new research capabilities: an acute, short-term demand for renewal posed by political and industrial stakeholders.

Harris' proposal for a "National Research and Resource Facility for Submicron Structures" (NRRFSS) was enthusiastically approved by his superiors, but the National Science Board (NSB; the NSF's governing body) was initially wary of awarding a block grant to a center rather than merit-based grants to individual investigators—a sign of how early Harris was in establishing center-based block grants as the new normal at NSF. The NSB also expressed anxiety that the NRRFSS might duplicate or compete with industrial efforts.[29] Yet supporters of the NRRFSS claimed that by fostering new forms of university–industry–government interaction the facility would aid, not hinder, an industry that was losing ground to global competition. As a report from a workshop Harris organized to gather support for his proposal put it:

> foreign competition [is] a subject not normally viewed as part of a NSF sponsored Workshop's concern. However, if the electrical engineering academic community is to assess its priorities for the 1980s, the health and vigor of the American electronics industry is an essential consideration.... When one considers the obituaries of such industries both here and abroad as consumer electronics, cameras, electron microscopes and large tankers

that have fallen before the intense developmental efforts that Japan has become properly and respectedly [*sic*] famous for, it raises a grim specter for this country. Other countries like Germany and France are active in sub-micron fabrication as well. We court serious economic danger if the United States government fails to respond with adequate resources in this new area for technological supremacy.[30]

Several NRRFSS bids highlighted industrial connections. The Lincoln Lab/MIT proposal, for instance, declared its facility's "intended purposes [would be] the development of submicrometer technology and the transfer of that technology to universities and commercial firms" and listed successful technology transfers from Lincoln Lab microfabrication research.[31] Similarly, a Penn/Drexel/Lehigh proposal played up those universities' proximity to (and endorsement from) major corporate R&D players such as IBM, Sperry-Univac, RCA, and Bell Labs, as well as "a small silicon house, MOS Technology of Valley Forge."[32] Cornell's team, too, told their dean that "we have a history of successful collaboration with industry in our semiconductor work. ([Harris'] Workshop felt industrial participation was important.)"[33]

The NRRFSS competition should therefore be seen in the context of the increasing importance of economic thinking in American science policy in the 1970s and the growing view that American universities should generate innovations (not just personnel) to flow into, and provide competitive advantage for, firms. Scholars of the 1970s economic turn in American science policy, particularly Elizabeth Popp Berman, have noted the use of University–Industry Research Centers as a new tool of technology transfer; in fact one of Berman's cases, the Silicon Structures Project at Caltech, was partly a microfabrication facility.[34]

A closer look at the NRRFSS, however, reveals details underplayed in studies such as Berman's. First, the specific woes of the American semiconductor industry—beyond just general economic malaise—were central to the emergence of academic centers, especially after 1975. Because it was such an obviously science-based industry, semiconductor manufacturing lent itself well to academic participation; microelectronics was the object of concerted national efforts by Japan and several Western European states which seemed to demand an energetic response from the American state; and microelectronics' importance to war-fighting lent that response an urgency not seen with respect to industries such as textiles and auto manufacturing.[35]

Second, the NRRFSS competition shows that the new university–industry centers of the 1970s were built on an earlier generation of centers linking universities and industry. The leading contenders for the NRRFSS were universities that possessed one of the MRLs formerly funded by the Defense Advanced Research Projects Agency (DARPA) because DARPA had, since 1964, insisted on greater "coupling" between the MRLs and industry.[36] The use of centers as points of contact between universities and industry may have increased since the mid-1970s, as has the economic justification for such centers; but the USA's reliance on such centers after 1975 was enabled by long-standing precedents.

2.4   THE FLOODGATES OPEN

Indeed, Cornell's status as lead center in the MRL program was probably decisive in its winning the NRRFSS competition. Cornell has hosted the USA's "national" academic microfabrication user facility ever since. The NRRFSS competition itself, however, spurred emergence of an organizational field of academic microfabrication user facilities far beyond Cornell. Initially, the leader of the Berkeley proposal, Tom Everhart, attempted to recruit the leader of the MIT/Lincoln Lab team, Hank Smith, to Berkeley with the offer that "You come out here as a faculty member. I'll raise the money, you do the work. We'll set up our own nanofabrication facility and we'll beat the pants off of Cornell."[37] As a counteroffer, the director of Smith's division at Lincoln Lab

> asked what I [Smith] was going to do. So I told him that I would like to demonstrate that the NSF had made a big mistake. He says, 'Great! Let's do it.' Just like that, they gave me a million dollars. A million dollar budget, where in the hell did that came from? I didn't know there was that much fat in the budget.[38]

Lincoln Lab's "Submicron Technology Program ... was operational by late 1977."[39] Meanwhile, MIT recruited Smith to build a Submicrometer Structures Laboratory on the main campus. That facility opened in 1978, before Cornell's NRRFSS did.[40] By 1980, Smith was a full-time faculty member at MIT.

The NRRFSS competition also triggered movement at Stanford's Integrated Circuits Laboratory (ICL), the leading academic facility that had *not* submitted a proposal. The director of the ICL, James Meindl, did not submit an NRRFSS bid because he was skeptical of the facility's long-term funding prospects and because he was unenthused by the administrative load of running a user facility. By 1978, though, Meindl and his department chair and ally, John Linvill, devised a "Center for Integrated Systems" (CIS) containing an expanded ICL containing shared equipment much like the NRRFSS and the Submicron Structures Laboratory.[41]

To fund the CIS, Linvill and Meindl carefully cultivated an elite group of American companies to form an "industrial affiliates" program.[42] They also struck up a vigorous correspondence with Jay Harris and his superiors, culminating in a visit to Stanford by James Krumhansl, Assistant Director for Mathematical and Physical Sciences and Engineering in October 1978.[43] Stanford's lobbying was fortuitous because Krumhansl was beginning to assemble an NSF program in "Microstructures Science, Engineering, and Technology." Krumhansl seems to have intended this program to foster a new proto-discipline focused on microstructures in the same way the Advanced Research Projects Agency (ARPA) used the MRLs to foster a new discipline of materials science in the early 1960s.

To aid that effort, Krumhansl was the "moving force" behind a workshop held in November 1978 at the NSF's Airlie House, organized by the Cornell facility and steered jointly by the NSF and a National Research Council (NRC) panel on Thin Film Microstructure Science and Technology.[44] The NRC panel framed its conclusions in the now-ubiquitous language conflating economic competition and national security:

> The United States has led in the development and exploitation of modern solid-state electronics technology; whether it will maintain this leadership is by no means certain.... Japanese industry, with active and extensive support from the Japanese Government, has mounted an intense research and development effort in microfabrication.... [S]ignificant research and development efforts are under way England, Holland, France, and West Germany. In addition, the technologies employed in national defense depend on semiconductor electronics; therefore, leadership in semiconductor electronics is essential to our national security.[45]

Remedying that would require "a new and expanded set of coordinated research programs in microstructure science and engineering," including "Regional Research Centers .... comparable in scope with the Materials Research Laboratories and the Cornell Submicron Facility."[46] The NSF was already imagining an organizational field of academic centers for what would become nanoscience research, nucleated around the NRRFSS and other microfabrication facilities.

Since the late 1970s, the Cornell, Stanford, and MIT facilities have vied for leadership of a microfabrication user facility organizational field populated by a growing list of peers. The University of Minnesota, for instance, formed a Microelectronics and Information Science Center (MISC) in 1980 to allow users access to "the processing facilities of nearby corporate contributors, including Control Data, Honeywell, Sperry, and 3 M."[47] Caltech did something similar with its Silicon Structures Project in 1977, and in 1981 Rensselaer Polytechnic followed with its Center for Integrated Electronics, and Arizona State with its Center for Solid State Electronics Research.[48] By the mid-1980s, one or two campus facilities opened every year at places like the Rochester Institute of Technology (1985), the University of Michigan (1986), Yale and the University of Cincinnati (both 1988).

In some cases, entrepreneurial faculty used the emergence of the academic microfabrication organizational field to attract resources to expand pre-existing rudimentary efforts. For instance, the University of Arkansas "obtained its first fabrication facilities in the late 1960's ... [but] in 1978, largely through the efforts of Dr. W.D. Brown ... the [EE] department obtained considerable additional equipment through grants from Sandia Laboratories, Texas Instruments, and the National Science Foundation."[49] Similarly, "development of the Auburn University Microelectronics Laboratory began in 1975," but "the Alabama Microelectronics Science and Technology Center (AMSTEC) was [only] formed at Auburn University in 1984, following a special legislative appropriation of $250,000/year."[50]

As the Auburn case implies, propagation of the microfabrication facility field sometimes occurred when state governments engaged in mesoscale (economic) competition with each other. The Microelectronics Center of North Carolina (MCNC), for instance, formed as a five-school state-funded consortium in 1981, championed by Governor James Hunt and funded with an initial $24.4 million from the state budget between 1981 and 1983. The Microelectronics Center of North Carolina (MCNC) was integral to North Carolina's success in attracting the Semiconductor Research Corporation, an industrial research consortium founded in 1982.[51]

Similarly, when the state of Texas wanted to woo another research consortium, the Microelectronics and Computer Technology Corporation (MCC), to Austin in 1983–84, part of Governor Mark White's pitch was that the state would put money into a new MRC at the University of Texas.[52] Then, when the state wanted to attract a semiconductor manufacturing R&D consortium, Sematech, in 1986–87, it built the MRC a brand-new, state-of-the-art facility.[53] These new microelectronics and microfabrication facilities almost always cited their predecessors as models and competitors. As the proposal for University of Texas' (UT) MRC put it in 1983:

> The economy of the State of Texas is rapidly moving toward high-technology industries, particularly in microelectronics and computers.... The purpose of this proposal is to insure that the University of Texas is the leader in that effort. Development of microelectronics research centers has begun at a number of universities (Table 1) as a response to the widely perceived necessity for fundamental and applied work in these areas.[54]

"Table 1" then listed, in order, data for the Cornell, Stanford, MIT, North Carolina, Arizona State, and Minnesota facilities. In other words, the mechanisms of "institutional isomorphism"—the mutual, active intercomparisons made among units in an organizational field leading to diffusion of norms across the field—were at work in the replication of the academic microfabrication user facility organizational form.[55]

2.5   Propaganda Value and Propagation of Values

For the leading microfabrication facilities, there were real benefits to fostering newly entrant peer facilities. As an MIT faculty member reported to the Submicrometer Structures Laboratory team after a visit to Stanford in 1977, "While it may seem strange to us, Jim Meindl said that he thought MIT's entry into the IC [integrated circuit] field would legitimize it, and give more emphasis to Stanford's program. I cannot overemphasize that everyone I met was *most* cordial and friendly, and eager to cooperate."[56] At Cornell, leaders of the National Submicron Facility advertised that they were a national resource not just for tools, but also for knowledge of how to establish and operate similar facilities. That knowledge aided propagation of the microfabrication organizational field, and helped maintain ties among organizations in that field. As a former director of the NRRFSS reported in 1986:

> GE had an engineer in residence at NRRFSS for a year .... [who] returned to GE and established a similar processing capability.... Strong interaction

continues between GE and Cornell. NRRFSS is continually called on to help/ advise other companies and universities in setting up similar laboratories, such as Varian, GE, McDonnell Douglas, the Jet Propulsion Lab, Hughes, Caltech, University of Michigan and University of California San Diego. Over the last several years we have advised more than forty organizations.[57]

NRRFSS's advisory role was actively fostered by the NSF:

A strong recommendation came out of the site review team that the facility host a meeting of microelectronics-related center directors to encourage collaborations and technology transfer. The NSF … has endorsed this concept and will both request the facility to do so and will provide funding for such a meeting. Coincidentally a Professor Marc Heritage from the University of Utah visited submicron the day after our site visit to discuss how to establish a similar center at the University of Utah.[58]

As the reference to "technology transfer" suggests, one point of organizational copying and competition was their university–industry partnerships, which took many forms: sharing facilities with industry; industry internships for students; annual corporate "membership" fees in return for previews of faculty research; and so on. Intelligence about industry partnerships diffused through: invitations to directors of other facilities to give presentations on their programs[59]; phone calls to industry leaders to ask how rival facilities approached their companies[60]; obtaining prospectuses for competitors' industrial programs[61]; and so on.

Intelligence about facilities' industrial partnerships and other practices also spread through the media. By 1985, the NRRFSS alone had appeared in some "22 magazines and 43 newspapers."[62] The NRRFSS also received ample opportunities from politicians for extolling its model for how American universities could better contribute to national economic competitiveness. For instance, in 1979 the House Subcommittee on Science, Research, and Technology summoned Ed Wolf, director of the facility, to testify on "Government and Innovation: University-Industry Relations." [63] Five years later, Cornell's president, Frank Rhodes, was also called before the House, where he pointed to the National Submicron Facility as an example of how to overcome the problem of access to increasingly expensive instrumentation.[64]

Yet despite favorable political and media attention, by the mid-1980s the NSF was not entirely happy with its flagship facility. This was, in part, a consequence of Cornell's *success* at building a local, cross-disciplinary stable of interconnected, federally funded centers. As a Cornell public relations officer put it in describing how the school acquired NSF funding for a supercomputing facility in 1985:

> We prepared a background piece saying that "Cornell University is a promising location for a national, advanced scientific computing center because of its experience in operating highly successful interdisciplinary centers for the benefit of the scientific research community." And we took the opportunity to brag about the Cornell Manufacturing Engineering and Productivity Program (COMEPP) and the Cornell High Energy Synchrotron Source (CHESS) and the Materials Science Center and the National Research and Resource Facility for Submicron Structures (which spells NRRFSS) and the Cornell Biotechnology Institute and the Semiconductor Research Corporation Center of Excellence in Microscience and Technology.[65]

Indeed, it was reported to Cornell's President Rhodes in 1986 that "there is growing concern at NSF, which may underlie [NSF Director] Erich Bloch's longstanding complaint about the Submicron Facility, that too much NSF money is going to New York State and particularly to Cornell."[66]

Ironically, the organizational model forged at Cornell contributed much to Bloch's own Engineering Research Centers (ERC) program, designed to address "immediate concerns in both engineering research and engineering education—concerns articulated by both academe and industry."[67] Over time, the ERCs spawned cascades of new center programs at the NSF: for example, the Science and Technology Centers and the Centers for Research Excellence in Science and Technology programs in 1987, the Materials Research Science and Engineering Centers (a revamped version of the MRLs) in 1994, and several smaller programs (e.g. Centers for Analysis and Synthesis; Centers for Chemical Innovation; Science of Learning Centers)—not to mention other centers that were not part of any larger center program. Centers—and especially programs spawning peer groups of centers—have become an almost instinctive mode of funding at the NSF and across American academia, government, and even industry.

These later center programs owed a great deal to the NRRFSS's example. As the industrial members of the NRRFSS' Policy Board argued in 1986, "The National Science Foundation's investment in the Submicron

Facility has enabled it to serve as a model for scientific and engineering centers nationally."[68] The then head of the Engineering Directorate, Nam Suh, similarly acknowledged that "the model of NRRFSS as a user facility role has been utilized in the planning and establishment of NSF's Regional Instrumental Laboratory program. More recently, the interdisciplinary operation of NRRFSS has provided the feasibility model for the innovative ERC program."[69]

This back and forth between Suh and the Cornell facility's board arose because the NRRFSS grant was ending in 1986. To keep going, Cornell proposed to turn the facility into an Engineering Research Center focused on nanoelectronics. This was bluntly turned down, leading to a scramble to exert industrial pressure on the NSF to save the facility. Ultimately, NSF relented and the NRRFSS was re-funded in 1987 as the "National Nanofabrication Facility" (NNF). Five years later, the NNF grant wound down and NSF made Cornell compete for to host the "national" facility.

This time, the Stanford ICL's director, James Plummer, eagerly stepped forward with a proposal. What happened next is still murky and subject to backroom gossip. It appears that, once again, Cornell's powerful supporters kept it from losing; but Stanford's equally powerful supporters and the originality of its proposal kept Cornell from winning outright. Caught in a bind, NSF withdrew the NNF competition and hastily announced a new contest for a National Nanofabrication Users Network (NNUN) of geographically distributed facilities. Initially, Cornell and MIT tried to pair up, but—probably responding to pressure from NSF—Cornell broke ranks and formed a consortium with Stanford, the University of California—Santa Barbara, Penn State, and Howard University (the USA's premier historically black university, located in Washington, DC).[70]

Ever since, that consortium—with Cornell as leader and Stanford as coleader—has thrived and grown. At the end of the NNUN grant's ten-year run, Cornell and Stanford competed and won against a consortium led by MIT and the University of Illinois for what was now known as the National Nanotechnology Infrastructure Network (NNIN). The NNIN was considerably larger than the NNUN, at more than a dozen facilities. As a result, the NNIN was also much more evenly distributed across the continental USA than the NNUN, which only covered the Mid-Atlantic and California.

2.6   CONCLUSION: INDUSTRY REORGANIZATION, RESEARCH REORIENTATION

The NNIN was also more disciplinarily diverse than the NNUN, and much more than the original NRRFSS and ICL. Although Hackett and Parker (this volume) are surely correct that formal, federally funded centers impose a rigid ideal of interdisciplinarity relative to the more free-form, emergent interdisciplinarity that occasionally arises organically from research communities, centers have the ability to provoke interdisciplinarity where it has not emerged organically—what the editors of this volume call "organizing interdisciplinary research."

For example, probably the most consequential innovation from the NRRFSS was an accidental (but not unintended) collaboration between electrical engineer Ed Wolf and horticulture professor John Sanford. Although Wolf and NSF had long promoted the NRRFSS as an interdisciplinary venture, for its first decade there were few projects centered on the life sciences. But when Sanford gave a talk around 1982 on unpromising attempts to drill holes in pollen grains with a microlaser—"with the purpose of letting DNA diffuse through the opening in the [pollen grain] wall"—one of his colleagues suggested the Submicron Facility might have more precise beams that could do the job better.[71] Eventually he found Wolf, and they came up with the messy, amusing, lucrative idea of blasting DNA-coated micron-scale tungsten particles into onion cells with an air gun. That idea generated what Nicole Nelson describes as "the largest royalty payment to the Cornell Research Foundation up to that date and ... one of the most 'readily recognized financial successes' in the history of Cornell technology transfer."[72]

Thus, centers can function as "trading zones" where practitioners from different backgrounds can orient to some common mission without fully understanding each other's knowledge, values, or techniques.[73] Ed Wolf and John Sanford would never have encountered each other without the NRRFSS' instigation to collaboration, even though their actual experiment involved little of the facility's equipment. The NRRFSS, as a new type of research facility, provided the opportunity structure for interdisciplinary encounters.

The NNIN's greater interdisciplinarity than the NRRFSS' is also due to changes in the technologies and structure of the microelectronics industry. There are more means available today for inspecting and manipulating nanoscale objects. The NRRFSS and its peers were founded in the "three beams" era, but today the three beams are joined by lithography and characterization with scanning tunneling microscopes and atomic force microscopes, flash imprint lithography, atom-probe tomography, and other techniques. As I have shown elsewhere, the proponents of those techniques have been skilled at interdisciplinary appeal.[74] Once any nanofabrication facility acquires, say, an atomic force microscope, it is instantly able to reach users in biology, polymer chemistry, geology, and other disciplines who were uninterested in microfabrication equipment in the late 1970s.

Conversely, the original industrial stakeholders of American microfabri-cation research are less central than they used to be. Most microelectronics manufacturing has moved across the Pacific, even if some of the biggest firms are still headquartered in the USA. Those firms, however, are no longer vertically integrated. When Stanford's CIS opened, its patrons

> were all vertically integrated companies. So, different pieces of different companies interfaced with different parts of the CIS structure.... It's dif-ferent today because many fewer companies are vertically integrated. Fabs are no longer commonplace in most companies, because of the existence of foundries. So the CIS model has had to evolve over time to recognize that many of the companies that belong to it are not vertically integrated and just connect in at specific points.[75]

The technology of microelectronics manufacturing has also evolved to diminish the centrality of academic research in the "three beams" mode. Against all odds, industry has stuck with optical lithography.[76] Academic advocates of e-beam, x-ray, and more exotic lithographies recognize their favored technique now has less chance of becoming an industrial mainstay than in 1975. Moreover, commercial manufacturing is so focused on mass production, and involves so many tightly interlinked process steps, that the gap between industrial lithography and its academic counterpart is wider than ever.

Microelectronics manufacturers are still important for academic micro-fabrication, of course, but less *sui generis* in their importance than during Cold War. But who now shares responsibility for patronage of the field? As it happens, the end of the Cold War saw an unprecedented hierar-chy reversal in American science, as federal funding for physical and engi-neering science declined while biomedical research skyrocketed. In that environment, savvy academic microfabrication specialists turned toward collaboration with life scientists—as did experts in many other fields such as synchrotron radiation physics.[77] Catherine Westfall has termed such pairings "recombinant science."[78] Declining support from traditional national security and microelectronics patrons meant microfabrication facilities depended upon users finding new patrons and applications and hence "recombining" their tools and expertise with unexpected partners.

Thus, 1982's accidental pairing of Wolf and Sanford became routine a few years later. For instance, Fabian Pease, one of Stanford's star electron beam lithography experts, spent the early 1990s using that technique to develop "gene chip" technology.[79] Similarly, the director of Cornell's NNF, Harold Craighead, organized "a highly successful and, in retrospect, pivotal workshop" on "Nanofabrication and Biosystems: Integrating Materials Science, Engineering, and Biology" in 1994 that staked his, and his facility's, claim in the field.[80] Six years later Craighead founded Cornell's Nanobiotechnology Center, which today is housed alongside (and uses the resources of) the NNF's successor facility. Papers with biological topics started to appear at meetings such as the Three Beams conference and the Gordon Conference series on Nanostructure Fabrication with greater frequency than a decade earlier. By 2013, five of the 14 NNIN facilities listed a life science area as a core field of expertise. Three (Georgia Tech, Washington University, and University of Washington) were predominantly biomedically oriented.

The diversification of applications in micro/nanofabrication was driven both by grassroots researchers responding to their environment and by top-down steering from federal agencies. For instance, when the NSF reviewed proposals for the NNUN in 1993, the *first two* questions it posed to principal investigators were:

> Specifically, what biologically-relevant projects will you target in future years?

> How do you propose to increase research in non-electronic areas such as condensed matter physics, materials, and chemistry through your proposed nanofabrication network?[81]

In the NNIN, each site carved a specialized niche, though each site also housed general-purpose equipment usable across many domains: Santa Barbara and Texas for compound semiconductors, Harvard handled the growing demand for information technology relevant to nanofabrication,

Penn State worked with a clean room technician training program, and so on. The NNIN's formal organizational field of nanofabrication facilities was characterized by the simultaneously centripetal and centrifugal forces typical of many postwar networks of research centers.[82]

That tension also governs the other formal and informal organizational fields that nanofabrication participates in. When they first emerged in the 1960s, academic microfabrication facilities were seen as university outposts of the microelectronics industry. Their success, however, turned center-based funding into a general tool of industrial and science policy. Now, every time the federal government identifies a new R&D objective, it funds a new system of centers or redirects an existing one. This was certainly the case when the National Nanotechnology Initiative (NNI) was formed in 2000. As of July 2012, the NNI's website listed 54 dedicated university nano centers funded by its member agencies, plus another six "networks" of smaller centers and 17 NSF-funded academic centers that are partially nano-oriented.

As federal science policy initiatives wax and wane, they leave behind systems of academic centers, each modeled on its predecessors and modeling for its successors. Over time, centers evolve and interconnect with other centers. Often, a university with one center can leverage acquisition of another. That has certainly been the case with microfabrication facilities. Of the 14 NNIN campuses, 12 also had a Materials Research Science and Engineering Center (MRSEC) and/or at least one other nano center (as categorized by the NNI). Five schools (including Cornell and Stanford) had an NNIN site, a MRSEC, plus *two* other NNI-defined nano centers.

If we want to understand the contemporary scientific enterprise, we need a better picture of how organizational fields of research centers form, operate, and evolve. This is true globally, across most scientific disciplines and high-tech industries, and across the university, government, and commercial sectors. This chapter has examined the particular organizational field of microfabrication facilities in American universities—a domain where systems of centers have had special salience. Academic microfabrication facilities merit such focused attention because they served as models both for new forms of university–industry–government interaction in the 1970s and 1980s and for later systems of federally funded academic centers. They were not, of course, the only such models, but in some fields—especially nanotechnology—these facilities so important as to be regarded as vital "infrastructure." In all likelihood, they will evolve into a new infrastructural role as policymakers' attention to nanotechnology wanes and is replaced by initiatives in energy and environment, synthetic biology, neuroscience, and so on.

NOTES

1. Gordon Research Conference, "Nanostructure Fabrication," http:// www.grc.org/programs.aspx?year=2012&program=nanofab (accessed July 15, 2013).
2. Post-academic: John Ziman, *Real Science: What It Is and What It Means* (Cambridge, UK: Cambridge University Press, 2000). Mode 1/Mode 2: Michael Gibbons et al., *The New Production of Knowledge: The Dynamics of Science and Research in Contemporary Societies* (London: Sage, 1994). Post-modern: Paul Forman, "The Primacy of Science in Modernity, of Technology in Postmodernity, and of Ideology in the History of Technology," *History and Technology* 23 (2007): 1–152. Neoliberal: Philip Mirowski, *Science-Mart: Privatizing American Science* (Cambridge, MA: Harvard University Press, 2011). Triple helix: Henry Etzkowitz, *The Triple Helix: University-Government-Industry Innovation in Action* (New York: Routledge, 2008).
3. Colin Norman, "Electronics Firms Plug into Universities," *Science* 217 (August 6, 1982): 511–514.
4. Paul DiMaggio and Walter W. Powell, "The Iron Cage Revisited: Institutional Isomorphism and Collective Rationality in Organizational Fields," *American Sociological Review* 48 (1983): 147–160.
5. See, for example, Roger Geiger, *To Advance Knowledge: The Growth of American Research Universities, 1900–1940* (New York: Oxford University Press, 1986); or the essays in *Osiris*, volume 20 (2005), "Politics and Science in Wartime: Comparative International Perspectives on the Kaiser Wilhelm Institute."
6. See, for example, the various extradepartmental laboratories established after the war at Stanford and MIT in Stuart W. Leslie, *The Cold War and American Science: The Military-Industrial-Academic Complex at MIT and Stanford* (New York: Columbia University Press, 1993).
7. Perhaps the best study that addresses this issue is Peter J. Westwick, *The National Labs: Science in an American System, 1947–1974* (Cambridge, MA: Harvard University Press, 2003). Recently, Hyungsub Choi and I have tried to raise this issue with respect to postwar systems of academic centers: Hyungsub Choi and Cyrus C.M. Mody, "From Materials Science to Nanotechnology: Institutions, Communities, and Disciplines at Cornell University, 1960–2000," *Historical Studies in the Natural Sciences* 43 (2012): 121–161.

8. Christophe Lécuyer, "Semiconductor Innovation and Entrepreneurship at Three University of California Campuses," in *Regional Economic Development, Public Universities, and Technology Transfer: Studies of the University of California's Contributions to Knowledge-based Growth,* ed. Martin Kenney and David Mowery (Stanford: Stanford University Press, forthcoming).

9. See "Back to the Future: Professor Dudley A. Buck (1927–1959)," *RLE Currents* 2.1 (December, 1988): 18–19, as well as oral history interview with Henry I. Smith conducted by the author, October 25, 2005, transcript available from the Chemical Heritage Foundation.

10. Mara Mills, "Hearing Aids and the History of Electronics Miniaturization," *IEEE Annals of the History of Computing* 33.2 (February 2011): 24–45.

11. Mark L. Schattenburg, "History of the 'Three Beams' Conference, the Birth of the Information Age, and the Era of Lithography Wars," 2007, http://eipbn.org/2010/wp-content/uploads/2010/01/EIPBN_history.pdf (accessed 15 Dec 2011).

12. D.R. Herriott, "The Development of Device Lithography," *Proceedings of the IEEE* 71.5 (1983): 566–570.

13. Christophe Lécuyer and David C. Brock, "From Nuclear Physics to Semiconductor Manufacturing: The Making of Ion Implantation," *History and Technology* 25.3 (2009): 193–217.

14. Cyrus C.M. Mody, "Conferences and the Emergence of Nanoscience," in *The Social Life of Nanotechnology,* ed. Barbara Herr Harthorn and John Mohr (London: Routledge, 2012): 52–65.

15. Interview with Dean Neikirk conducted by the author, October 12, 2009, Austin, TX.

16. Oral history interview with David A. Hodges conducted by Christophe Lécuyer, between June 2008 and June 2010, available at http://andros.eecs.berkeley.edu/~hodges/David_Hodges_Interviews.pdf.

17. *An Assessment of the Needs for Equipment, Instrumentation, and Facilities for University Research in Science and Engineering* (Washington, DC: National Academy of Sciences, 1971).

18. W. Patrick McCray, *Giant Telescopes: Astronomical Ambition and the Promise of Technology* (Cambridge, MA: Harvard University Press, 2004).

19. Olof Hallonsten, "Small Science on Big Machines: Politics and Practices of Synchrotron Radiation Laboratories" (PhD dissertation, Lund University, 2009).

20. Proposal for "Operational support of the regular collection of mammals in the Museum of Vertebrate Zoology," W.Z. Lidicker and J.L. Patton, November 23, 1977, Museum of Vertebrate Zoology archives (courtesy of Mary Sunderland).

21. William J. Cromie, "Regional Instrumentation Centers," *Mosaic* 11.2 (1980), 12–18.

22. Dian Belanger, *Enabling American Innovation: Engineering and the National Science Foundation* (West Lafayette, IN: Purdue University Press, 1998).

23. Jay Harris, "It's a Small World," text of talk delivered at the 25th anniversary of the Cornell Nanofabrication Facility/NRRFSS in 2003, copy given to the author by Jay Harris.

24. Arthur Fisher, "The Magnetism of a Shared Facility," *Mosaic* 11.2 (1980), 38–45. Jay Harris claims that he had the FBNML in mind when he conceived the idea for a national microfabrication facility. Jay Harris, interview with author, San Diego, CA, May 5, 2006.

25. Charles Polk, "Address to the NSF Workshop," in *Report of the NSF Workshop on Needs for a National Research and Resource Center in Submicron Structures (East Coast),* [henceforth Penn workshop report] submitted by J.N. Zemel and M.S. Chang (Philadelphia: Moore School of Electrical Engineering, University of Pennsylvania, May 10, 1976), 25–35. Emphasis in original.

26. Harris, "Small World."

27. E.D. Wolf and J.M. Ballantyne, "Research and Resource at the National Submicron Facility," in Norman G. Einspruch, ed., *VLSI Electronics: Microstructure Science, vol. 1,* (New York: Academic Press, 1981), 129–183.

28. Scott Callon, *Divided Sun: MITI and the Break-Down of Japanese High-Tech Industrial Policy, 1975–1993* (Stanford: Stanford University Press, 1995).

29. Jay H. Harris et al., "The Government Role in VLSI," in Norman G. Einspruch, ed., *VLSI Electronics: Microstructure Science, vol. 1* (New York: Academic Press, 1981), 265–299.

30. Zemel and Chang, *Report of the [Penn] Workshop,* p. 49–50; see also *Needs for a National Research and Resource Center in Submicron Structures: Report on National Science Foundation Workshop Held in Salt Lake City, Utah May 21, 1976,* submitted by Richard W. Grow, Robert J. Huber, and Roland W. Ure, Jr. (Salt Lake City: Microwave Device and Physical Electronics Laboratory, University of Utah, 15 Sept, 1976), pp. 106–7.

31. Henry Smith (PI), "Proposal submitted to National Science Foundation for National Research and Resource Facility for Submicron Structures," July 1, 1977; copy given to the author by Henry Smith.

32. Jay N. Zemel (PI), "Proposal for a National Center for Submicron Structure Research" [University of Pennsylvania/Drexel University/ Lehigh University], January 1977. Copy given to author by Ed Wolf.

33. "Some reasons why Cornell has a strong chance to attract such a center," undated but probably summer, 1976, probably from Joe Ballantyne and/or Charles Lee, in Cornell Center for Materials Research [henceforth CCMR] records, #53-24-3676, Box 29, Folder 37, Division of Rare and Manuscript Collections, Cornell University Library.
34. Elizabeth Popp Berman, *Creating the Market University: How Academic Science Became an Economic Engine* (Princeton: Princeton University Press, 2012).
35. All three rationale are laid out in, among many others, the "Report of the White House Science Council Panel on Semiconductors," September 1987, William Graham files collection, Box CFOA 990, Folder "Semiconductors (1)," Ronald Reagan Presidential Library.
36. This point is emphasized in Mody and Choi, 2013.
37. Smith oral history.
38. *Ibid.*
39. MIT Research Laboratory for Electronics, *Currents* 2.1 (December 1988): 10; also Smith oral history and *Report of the President and Chancellor 1977–78* (Cambridge, MA: Massachusetts Institute of Technology, 1978).
40. *Ibid.* (all references). Also, *Report of the President and Chancellor 1978–79* (Cambridge, MA: Massachusetts Institute of Technology, 1979).
41. Forest Baskett et al. (including Linvill, Meindl, and James Gibbons), Proposal to Establish the Center for Integrated Systems at Stanford University, August 1979, CIS in-house archive.
42. Kevin Gross, "CIS Groundbreakers Laud Research-Industry Ties," *The Stanford Daily*, May 20, 1983, in CIS in-house archive.
43. I have found a dozen letters between Stanford faculty (Linvill, Meindl, and Bill Spicer) and Harris, Chenette, Bourne, and Krumhansl between June 27 and September 29, 1978 in the Center for Integrated Systems in-house archive. The CIS archive also contains a number of letters written in the same period that circulated either within Stanford or within federal science policy circles discussing plans for the CIS.
44. Joseph Ballantyne, "Introduction," *Proceedings: NSF Workshop on Opportunities for Microstructures Science, Engineering and Technology in cooperation with the NRC Panel on Thin Film Microstructure Science and Technology : November 19–22, 1978* (Washington, DC: National Science Foundation, 1978).
45. National Research Council panel on Thin-Film Microstructure Science and Technology, *Microstructure Science, Engineering, and Technology* (Washington, DC: National Academy of Sciences, 1979), 1.
46. *Ibid.*
47. Dale Whittington, *High Hopes for High Tech: Microelectronics Policy in North Carolina* (Chapel Hill: University of North Carolina Press, 1985),

138. Also Gregory T. Cibuzar, "Microelectronics at the University of Minnesota," *Proceedings of the Tenth Biennial University/Government/ Industry Microelectronics Symposium* (IEEE: 1993): 170–173.

48. John F. Mason, "VLSI Goes to School," *IEEE Spectrum* (November 1980): 48–52.

49. Jerry R. Yeargan, "Developing a Program in Analog Electronics," *Proceedings of the Sixth Biennial University/Government/Industry Microelectronics Symposium* (IEEE: 1985): 9–11; Robert M. Burger, *Cooperative Research: The New Paradigm* (Durham, NC: Semiconductor Research Corporation, 2001). Note that the SRC operates largely through project grants to universities, and that its first University Advisory Committee was composed of Ed Wolf, John Linvill, Ben Streetman from Illinois (but who would soon found Texas' MRC), Paul Penfield (associated with the MIT Submicrometer Structures Laboratory), faculty members associated with the Arizona State, Caltech, and Minnesota facilities listed above, plus professors from Berkeley and Carnegie-Mellon.

50. R.C. Jaeger, et al., "The Alabama Microelectronics Science and Technology Center and the Microelectronics Program at Auburn University," *Proceedings of the Sixth Biennial University/Government/Industry Microelectronics Symposium* (IEEE: 1985): 42–45.

51. *High Hopes for High Tech*, 14; H. Craig Casey, "The Microelectronics Center of North Carolina: A New Program in Science and Integrated Circuits," *Proceedings of the Fourth Biennial University/Government/ Industry Microelectronics Symposium* (IEEE: 1981): 169 ff.

52. University of Texas Office of Public Affairs, news release for July 12, 1984, in UT Office of Public Affairs Records, Dolph Briscoe Center for American History, University of Texas at Austin, Subject Files, Box 4Ac86, Folder "Microelectronics Research Center."

53. University of Texas Office of Public Affairs, news release for April 9, 1987, in UT Office of Public Affairs Records, Dolph Briscoe Center for American History, University of Texas at Austin, Subject Files, Box 4Ac86, Folder "Microelectronics Research Center."

54. Ben Streetman et al., "Proposal for a Texas Microelectronics Center in the College of Engineering, The University of Texas at Austin," March 1983, in University of Texas Executive Vice President and Provost's Office records (collection 96–273), Dolph Briscoe Center for American History, University of Texas at Austin, Box 24, Folder Microelectronics Research Center, 1983–84.

55. For discussion of institutional isomorphism, see the essays in Walter W. Powell and Paul J. DiMaggio (eds.), *The New Institutionalism in Organizational Analysis* (Chicago: University of Chicago Press, 1991).

56. Letter from Prof. Peter A. Wolff to Prof. J. Allen, August 23, 1977, re: JSEP Topical Review of Semi-conductor Integrated Circuits, Devices, and Materials, Stanford, 3, 4 August 1977, Jerome Wiesner administration records, collection AC008, Box 205, Folder Research Laboratory for Electronics, MIT Institute Archives.

57. Joe Ballantyne, memo to Greg Galvin, re: Information for Frank Rhodes [president of Cornell], August 11, 1986, Frank H.T. Rhodes papers [henceforth FRP], #3-12-1795, Division of Rare and Manuscript Collections, Cornell University Library, Box 160, Folder 59.

58. NRRFSS Policy Board, "Background material for topics to be discussed," 17 Nov 1986, in DOR, Box 39, Folder 13.

59. Streetman, "Proposal," appendix reporting on talks by microfabrication facility directors from (in order or presentation) Stanford, Caltech, Cornell, RPI, Minnesota, North Carolina, Arizona State, Berkeley, and MIT.

60. I found a description of such a phone call in John Young (Hewlett-Packard), letter to Bruce Hinchcliffe, July 14, 1981, in Stanford Center for Integrated Systems in-house archive.

61. MIT Microsystems Industrial Group Prospectus, May 1989, found in Stanford CIS "archive".

62. NRRFSS, "Report on the National User Research Program '82-'85," in DOR, Box 38, Folder 5; also E.D. Wolf, "Recent National Press Coverage of NRRFSS," handout for NSF discussions, 7 Jan 1982, DOR, Box 37, Folder 32.

63. Richard Atkinson, testimony before the House Committee on Science and Technology, Subcommittee on Science, Research, and Technology hearing on Government and Innovation: University-Industry Relations, July 31, 1979, p. 82.

64. Frank H.T. Rhodes, testimony before the House Committee on Science and Technology, hearing on Improving the Research Infrastructure at US Universities and Colleges, May 8, 1984, p. 88.

65. Roger Segelken, "A Case History of a Computer Media Event—Introducing a Supercomputer Center," *Proceeding of the Fourth International Conference on Systems Documentation* (New York: ACM, 1985), 146–160.

66. John F. Burness, Vice President for University Relations, to Frank Rhodes, President of Cornell, September 10, 1986, FRP, Box 160, Folder 59.

67. Belanger, *Enabling American Innovation*, 219.

68. Submicron Facility Policy Board letter, October 16, 1986.

69. Nam Suh, NSF Assistant Director for Engineering, to National Science Board Committee on Programs and Plans, around October 16, 1986, re: Report on User Research at the National Research and Resource Facility for Submicron Structures, in FRP, Box 160, Folder 59.

70. Both Stanford's 1992 proposal for the National Nanofabrication Users Facility competition and documents relating to Stanford and Cornell's 1993 bid for the National Nanofabrication Users Facility *Network* competition were given to the author by Mary Tang and Nancy Latta.

71. John C. Sanford, "The Development of the Biolistic Process," *In Vitro Cellular and Developmental Biology—Plant* 36.5 (2000): 303–308.

72. Nicole Nelson, "Shooting Genes, Distributing Credit: Narrating the Development of the Biolistic Gene Gun," *Science as Culture* 21.2 (2012): 205–232.

73. Peter Galison, *Image and Logic* (Chicago: University of Chicago Press, 1997); Harry Collins, Robert Evans, Mike Gorman, "Trading Zones and Interactional Expertise," *Studies in History and Philosophy of Science Part A* 38 (2007): 657–666.

74. Cyrus C.M. Mody, *Instrumental Community: Probe Microscopy and the Path to Nanotechnology* (Cambridge, MA: MIT Press, 2011).

75. Interview with James Plummer conducted by the author, August 11, 2010, Palo Alto, CA.

76. Rebecca Henderson, "Of Life Cycles Real and Imaginary: The Unexpectedly Long Old Age of Optical Lithography," *Research Policy* 24 (1995): 631–643.

77. See Hallonsten and Heinze (this volume) as well as Park Doing, *Velvet Revolution at the Synchrotron: Biology, Physics, and Change in Science* (Cambridge, MA: MIT Press, 2009).

78. Catherine Westfall, "Rethinking Big Science: Modest, Mezzo, Grand Science and the Development of the Bevalac, 1971–1993," *Isis* 94 (2003): 30–56.

79. Doogab Yi, The Integrated Circuit for Bioinformatics: The DNA Chip and Materials Innovation, white paper for Studies in Material Innovation series (Philadelphia: Chemical Heritage Foundation, 2010) and Tim Lenoir and Eric Giannella, "The Emergence and Diffusion of DNA Microarray Technology," *Journal of Biomedical Discovery and Collaboration* 1 (2006).

80. Edward Wolf, "A Personalized Summary of CNF History," in *The Future of Nanotechnology* (Ithaca, NY: Cornell NanoScale Science and Technology Facility, 2007).

81. Linton G. Salmon (program director, Solid State and Microstructures), Adriaan M. de Graaf (deputy director, Division of Materials Research), and Michael K. Lamvik (program director, Instrumentation and Instrument Development), National Science Foundation, to James Plummer, July 20, 1993, packet of materials in preparation for site visit review. Provided to the author by Mary Tang and Nancy Latta.

82. See Westwick, *The National Labs* for a similar dynamic—each lab in the National Lab system wanted to align and compete with the others in shared domains, but also carve out its own domain distinctive from the rest.

REFERENCES

Atkinson, Richard. 1979. Testimony before the House Committee on Science and Technology, Subcommittee on Science, Research, and Technology hearing on Government and Innovation: University-Industry Relations, 31 July 1979.

Back to the future: Professor Dudley A. Buck (1927–1959). 1988. *RLE Currents* 2.1: 18–19.

Ballantyne, Joseph. 1978. Introduction. In *Proceedings: NSF workshop on opportunities for microstructures science, engineering and technology in cooperation with the NRC panel on thin film microstructure science and technology: November 19–22, 1978*. Washington, DC: National Science Foundation.

Ballantyne, Joe. 1986. Memo to Greg Galvin, re: Information for Frank Rhodes [president of Cornell], August 11, 1986, Frank H.T. Rhodes papers [henceforth FRP], #3-12-1795, Division of Rare and Manuscript Collections, Cornell University Library, Box 160, Folder 59.

Baskett, Forest, Meindl Linvill, and James Gibbons. 1979. Proposal to establish the Center for Integrated Systems at Stanford University, August 1979, CIS in-house archive.

Belanger, Dian. 1998. *Enabling American innovation: Engineering and the National Science Foundation*. West Lafayette: Purdue University Press.

Burger, Robert M. 2001. *Cooperative research: The new paradigm*. Durham: Semiconductor Research Corporation.

Burness, John F. 1986. Letter to Frank Rhodes, President of Cornell, 10 Sept 1986, FRP, Box 160, Folder 59.

Callon, Scott. 1995. *Divided Sun: MITI and the break-down of Japanese high-tech industrial policy, 1975–1993*. Stanford: Stanford University Press.

Casey, H. Craig. 1981. The Microelectronics Center of North Carolina: A new program in science and integrated circuits. In *Proceedings of the Fourth Biennial University/Government/Industry Microelectronics Symposium*. New York: New York.

Cibuzar, Gregory T. 1993. Microelectronics at the University of Minnesota. In *Proceedings of the Tenth Biennial University/Government/Industry Microelectronics Symposium*. New York: New York.

Collins, Harry, Robert Evans, and Mike Gorman. 2007. Trading zones and interactional expertise. *Studies in History and Philosophy of Science Part A* 38: 657–666.

Cromie, William J. 1980. Regional instrumentation centers. *Mosaic* 11(2): 12–18.

DiMaggio, Paul, and Walter W. Powell. 1983. The iron cage revisited: Institutional isomorphism and collective rationality in organizational fields. *American Sociological Review* 48: 147–160.

Doing, Park. 2009. *Velvet revolution at the synchrotron: Biology, physics, and change in science.* Cambridge, MA: MIT Press.

Etzkowitz, Henry. 2008. *The triple helix: University-government-industry innovation in action.* New York: Routledge.

Fisher, Arthur. 1980. The magnetism of a shared facility. *Mosaic* 11(2): 3845.

Forman, Paul. 2007. The primacy of science in modernity, of technology in postmodernity, and of ideology in the history of technology. *History and Technology* 23: 1–152.

Galison, Peter. 1997. *Image and logic.* Chicago: University of Chicago Press.

Geiger, Roger. 1986. *To advance knowledge: The growth of American Research Universities, 1900–1940.* New York: Oxford University Press.

Gibbons, Michael, and Peter Scott. 1994. *The new production of knowledge: The dynamics of science and research in contemporary societies.* London: Sage.

Gordon Research Conference. 2012. Nanostructure fabrication. http://www.grc. org/programs.aspx?year=2012&program=nanofab. Accessed 15 July 2013.

Gross, Kevin. 1983. CIS groundbreakers laud research-industry ties. *The Stanford Daily,* May 20.

Grow, Richard W., Robert J. Huber, and Roland W. Ure Jr. 1976. *Needs for a National Research and Resource Center in submicron structures: Report on National Science Foundation workshop held in Salt Lake City, Utah May 21, 1976.* Salt Lake City: Microwave Device and Physical Electronics Laboratory, University of Utah.

Hallonsten, Olof. 2009. Small science on big machines: Politics and practices of synchrotron radiation laboratories. PhD dissertation, Lund University.

Harris, Jay. 2003. It's a small world. Talk delivered at the 25th anniversary of the Cornell Nanofabrication Facility/NRRFSS.

Harris, Jay, et al. 1981. The government role in VLSI. In *VLSI electronics: Microstructure science,* vol. 1, ed. Norman G. Einspruch, 265–299. New York: Academic.

Henderson, Rebecca. 1995. Of life cycles real and imaginary: The unexpectedly long old age of optical lithography. *Research Policy* 24: 631–643.

Herriott, D.R. 1983. The development of device lithography. *Proceedings of the IEEE* 71(5): 566–570.

Jaeger, R.C., et al. 1985. The Alabama Microelectronics Science and Technology Center and the microelectronics program at Auburn University. In *Proceedings of the Sixth Biennial University/Government/Industry Microelectronics Symposium.* New Jersey: Piscataway.

Lécuyer, Christophe. forthcoming. Semiconductor innovation and entrepreneurship at three University of California campuses. In *Regional economic* development, public universities, and technology transfer: Studies of the University of California's contributions to knowledge-based growth, ed. Martin Kenney and David Mowery. Stanford: Stanford University Press.

Lécuyer, Christophe. 2012. Oral history interview with David A. Hodges, between June 2008 and June 2010, available at http://andros.eecs.berkeley.edu/~hodges/David_Hodges_Interviews.pdf

Lécuyer, Christophe, and David C. Brock. 2009. From nuclear physics to semiconductor manufacturing: The making of ion implantation. History and Technology 25(3): 193–217.

Lenoir, Tim, and Eric Giannella. 2006. The emergence and diffusion of DNA microarray technology. Journal of Biomedical Discovery and Collaboration 1: 11.

Leslie, Stuart W. 1993. The Cold War and American science: The military-industrial-academic complex at MIT and Stanford. New York: Columbia University Press.

Lidicker, W.Z., and J.L. Patton. 1977. Proposal for "Operational support of the regular collection of mammals in the Museum of Vertebrate Zoology." 23 November 1977, Museum of Vertebrate Zoology archives (courtesy of Mary Sunderland).

Mason, John F. 1980. VLSI goes to school. IEEE Spectrum 17(11): 48–52.

McCray, W. Patrick. 2004. Giant telescopes: Astronomical ambition and the promise of technology. Cambridge, MA: Harvard University Press.

Mills, Mara. 2011. Hearing aids and the history of electronics miniaturization. IEEE Annals of the History of Computing 33(2): 24–45.

Mirowski, Philip. 2011. Science-mart: Privatizing American science. Cambridge, MA: Harvard University Press.

MIT. 1988. Research Laboratory for Electronics. Currents 2(1): 1–15.

Mody, Cyrus C.M. 2011. Instrumental community: Probe microscopy and the path to nanotechnology. Cambridge, MA: MIT Press.

Mody, Cyrus C.M. 2012. Conferences and the emergence of nanoscience. In The social life of nanotechnology, ed. Barbara Herr Harthorn and John Mohr, 52–65. London: Routledge.

Mody, Cyrus C.M., and Hyungsub Choi. 2012. From materials science to nanotechnology: Institutions, communities, and disciplines at Cornell University, 1960–2000. Historical Studies in the Natural Sciences 43: 121–161.

National Academy of Sciences. 1971. An assessment of the needs for equipment, instrumentation, and facilities for university research in science and engineering. Washington, DC: National Academy of Sciences.

National Research Council panel on Thin-Film Microstructure Science and Technology. 1979. Microstructure science, engineering, and technology. Washington, DC: National Academy of Sciences.

Nelson, Nicole. 2012. Shooting genes, distributing credit: Narrating the development of the biolistic gene gun. Science as Culture 21(2): 205–232.

Norman, Colin. 1982. Electronics firms plug into universities. *Science* 217: 511–514.

NRRFSS. 1985. Report on the National User Research Program '82–'85. In DOR, Box 38, Folder 5.

NRRFSS Policy Board. 1986. Background material for topics to be discussed, 17 Nov 1986. In DOR, Box 39, Folder 13.

Polk, Charles. 1976. Address to the NSF workshop. In *Report of the NSF workshop on needs for a National Research and Resource Center in submicron structures (East Coast)*, ed. J.N. Zemel and M.S. Chang, 25–35. Philadelphia: Moore School of Electrical Engineering, University of Pennsylvania.

Popp Berman, Elizabeth. 2012. *Creating the market university: How academic science became an economic engine*. Princeton: Princeton University Press.

Powell, Walter W., and Paul J. DiMaggio. 1991. *The new institutionalism in organizational analysis*. Chicago: University of Chicago Press.

Report of the President and Chancellor 1977–78. 1978. Cambridge, MA: Massachusetts Institute of Technology.

Report of the President and Chancellor 1978–79. 1979. Cambridge, MA: Massachusetts Institute of Technology.

Rhodes, Frank H.T. 1984. Testimony before the House Committee on Science and Technology, hearing on improving the research infrastructure at US universities and colleges, 8 May 1984.

Sanford, John C. 2000. The development of the biolistic process. *In Vitro Cellular and Developmental Biology—Plant* 36(5): 303–308.

Schattenburg, Mark L. History of the 'Three Beams' conference, the birth of the information age, and the era of lithography wars. http://eipbn.org/2010/wp-content/uploads/2010/01/EIPBN_history.pdf. Accessed 15 Dec 2011.

Segelken, Roger. 1985. A case history of a computer media event: Introducing a supercomputer center. In *Proceeding of the Fourth International Conference on Systems Documentation*, 146–160. New York: New York.

Smith, Henry. 1977. *Proposal submitted to National Science Foundation for National Research and Resource Facility for Submicron Structures*. Lexington: MIT Lincoln Lab.

Streetman, Ben, et al. 1983–1984. Proposal for a Texas Microelectronics Center in the College of Engineering, The University of Texas at Austin. University of Texas Executive Vice President and Provost's Office records (collection 96–273), Dolph Briscoe Center for American History, University of Texas at Austin, Box 24, Folder Microelectronics Research Center.

Suh, Nam. Letter to National Science Board Committee on Programs and Plans, around 16 Oct 1986, re: Report on user research at the National Research and Resource Facility for Submicron Structures, in FRP, Box 160, Folder 59.

University of Texas Office of Public Affairs. 1984. News release for July 12, 1984, in UT Office of Public Affairs Records, Dolph Briscoe Center for American

History, University of Texas at Austin, Subject Files, Box 4Ac86, Folder "Microelectronics Research Center."

University of Texas Office of Public Affairs. 1987. News release for April 9, 1987, in UT Office of Public Affairs Records, Dolph Briscoe Center for American History, University of Texas at Austin, Subject Files, Box 4Ac86, Folder "Microelectronics Research Center."

Westfall, Catherine. 2003. Rethinking big science: Modest, mezzo, grand science and the development of the Bevalac, 1971–1993. *Isis* 94: 30–56.

Westwick, Peter J. 2003. *The National Labs: Science in an American system, 1947–1974.* Cambridge, MA: Harvard University Press.

Whittington, Dale. 1985. *High hopes for high tech: Microelectronics policy in North Carolina.* Chapel Hill: University of North Carolina Press.

Wolf, E.D. 1982. Recent national press coverage of NRRFSS. Handout for NSF discussions, 7 Jan 1982, DOR, Box 37, Folder 32.

Wolf, Edward. 2007. A personalized summary of CNF history. In *The future of nanotechnology.* Ithaca: Cornell NanoScale Science and Technology Facility.

Wolf, E.D., and J.M. Ballantyne. 1981. Research and resource at the National Submicron Facility. In *VLSI electronics: Microstructure science,* vol. 1, ed. Norman G. Einspruch, 129–183. New York: Academic.

Wolff, Peter A. 1977. Letter to Prof. J. Allen, 23 Aug 1977, re: JSEP Topical Review of Semi-conductor Integrated Circuits, Devices, and Materials, Stanford, 3, 4 Aug 1977. Jerome Wiesner administration records, collection AC008, Box 205, Folder Research Laboratory for Electronics, MIT Institute Archives.

Yeargan, Jerry R. 1985. Developing a program in analog electronics. *Proceedings of the Sixth Biennial University/Government/Industry Microelectronics Symposium,* 9–11. New Jersey: Piscataway.

Yi, Doogab. 2010. *The integrated circuit for bioinformatics: The DNA chip and materials innovation, white paper for studies in material innovation series.* Philadelphia: Chemical Heritage Foundation.

Young, John. 1981. Letter to Bruce Hinchcliffe, 14 July 1981. In Stanford Center for Integrated Systems in-house archive.

Zemel, Jay N. 1977. Proposal for a National Center for Submicron Structure Research. University of Pennsylvania/Drexel University/Lehigh University, January 1977. Pennsylvania: Philadelphia.

Ziman, John. 2000. *Real science: What it is and what it means.* Cambridge, UK: Cambridge University Press.

CHAPTER 3

From Salomon's House to Synthesis Centers

Edward J. Hackett and John N. Parker

3.1   INTRODUCTION

To meet the emerging challenges of the day, science has continually undergone a process of renewal that has extended beyond theories, results, and research technologies to include innovations in the organizational arrangements, collaborative dynamics, and epistemic principles that generate, distribute, and institutionalize new knowledge and know-how. Scientific synthesis centers are recent innovations in this process of renewal and examples of what the editors of this book call "investments in exploration" and "organization of interdisciplinary research." These innovative organizations catalyze and host working groups that integrate scientific diversity and engage real-world problems. In this chapter, we examine how synthesis centers arose through the interaction of intellectual and organizational innovations, then we will use ideas from science studies, small group dynamics, and the creativity and interdisciplinarity literatures

E.J. Hackett (✉)
Brandeis University, Waltham, MA, USA

J.N. Parker
Arizona State University, Tempe, AZ, USA

© The Editor(s) (if applicable) and The Author(s) 2016
T. Heinze, R. Münch (eds.), *Innovation in Science and Organizational Renewal*, DOI 10.1057/978-1-137-59420-4_3

to identify the patterns and processes of social interaction responsible for the center's performance. The processes of transformation within and outside the center are still at work, and so we will close with some observations about likely directions of change.

3.2   LOOKING BACKWARD

Science is a grand experiment, both in the conduct of inquiry and in the arrangement of institutions, organizations, and groups that structure the work and work lives of scientists. People have long pondered the optimal arrangement of science for producing sound and original knowledge that also contributes to human well-being. The scientific community of Renaissance Italy (ca. 1400–1500), for example, brought together scientists and scholars, humanists and tradesmen in what we today would call transdisciplinary collaborations (see editor's introduction). But there were not organizational and institutional bonds strong enough to hold such collaborations together, and so they fell apart. Patterns of organization that succeeded at certain times, in certain places, and for certain purposes may be inadequate in other circumstances. Looking backward at pioneering experiments in the organization of scientific work is a useful starting point for orienting and guiding our thinking. What can we learn?

The early seventeenth century offers several noble efforts to shape science to social purposes through innovative research organizations. Of these, Salomon's House is perhaps the best known (in English) vision of interdisciplinary inquiry organized for societal benefit, inspiring for centuries the formation of scientific associations and research organizations.[1] And "[since] the beginning of modern science and knowledge production is generally associated with Francis Bacon," this is a suitable point of departure.[2] In Bacon's conception, Salomon's House is a brotherhood of science and the useful arts embedded deep within the utopian society of Bensalem, dedicated to "the knowledge of causes, and secret motions of things; and the enlarging of the bounds of human empire, to the effecting of all things possible."[3] Among the research capabilities of Salomon's House are caves for refrigeration and simulation of cold environments, towers for observation, artificial wells and pools (some infused with various substances), experimental plots and populations, a place where life processes are sustained after death, an operating theater, brew houses, bake houses, medicine shops, machine shops, furnaces, a gem collection, optics and acoustics labs, a weapons lab, a computer (powered by people), and a simulation cave. Bacon prepared a detailed staffing plan for Salomon's House with job titles that included Merchants of Light (who conduct literature searches by ship), Depredators, Mystery-men, Pioneers or Miners (who try experiments), Compilers, Benefactors, Lamps, Inoculators, and Interpreters of Nature (who synthesize research results into laws and theories).

Bacon's vision is comprehensive and detailed, laudatory for its dedication of science to social ends, including education and public understanding. But public engagement—any sort of engagement with civil authority or society—is notably restricted: "And this we do also: we have consultations, which of the inventions and experiences which we have discovered shall be published, and which not: and take all an oath of secrecy, for the concealing of those which we think fit to keep secret: though some of those we do reveal sometimes to the state and some not."[4] As one commentator notes, "Bacon's science is rooted in the natural environment, not in the social structure....we can also read the arrangement of these structures to say that so far as science is creative discovery, it cannot flourish within the shadow of places devoted to other purposes."[5]

Salomon's House may be the best known utopian vision of science to emerge from the seventeenth century, but it is not the only one. Where Bacon set science apart from the social order and entrusted it with power over the direction, disclosure, and applications of scientific knowledge, Tomasso Campanella's *City of the Sun* (1602) and Johann Valentin Andreae's *Christianopolis* (1619)[6] imagined a scientific institution integrated with society and charged with deeper responsibility for human well-being. Frederico Cesi, a young Roman nobleman, went considerably further in the depth, scope, and enactment of his vision when, in 1603 and with the assistance of three young friends, he founded the Accademia dei Lincei (the Academy of the Lynx, likening scientists' perceptiveness to the sharp-eyed lynx) "to promote, coordinate, integrate, and spread scientific knowledge in its highest expressions...not only to acquire knowledge and wisdom for living righteously and piously, but with voice and writing reveal them unto men."[7] His expansive and integrative vision is global, encompasses the world of scholarship, and prominently positions values, principles, and public enlightenment in the scientific Accademia's design. Membership was open to all scientists and to humanists, but clergy were unwelcome. A main facility in Rome was planned, complete with library and laboratories, machinery, optical equipment, scientific instruments, a printing office, museum, and botanical garden. Lesser facilities were to be constructed in the four quarters of the world. The vision is magnificent and enticing: an interdisciplinary scientific organization of global ambition equipped with the latest technology and devoted to advancing human well-being and public understanding of science. Galileo joined in 1611. But the early history of the Accademia was troubled and tragic, as it was attacked by the church and clouded by the founder's premature death. But the Accademia was revived and endures, and retains its home in the Palazzo Corsini and Villa Farnesina.

Each pioneer crafted an arrangement for the creation, diffusion, and utilization of scientific knowledge that suited the circumstances of his day. Applications of science to promote wealth and well-being predominate, along with commitments to basic research. Bacon was so intent on freeing science from a potentially oppressive social order that he sequestered its activities, placing them in the hands of a benevolent but secretive brotherhood. Cesi's vision might be the most complete and endearing—and so powerful that it aroused powerful opposition.

Much has changed since these visionaries imagined ways to organize science: discovery has made the world larger, while technology has made it smaller. Nation-states formed, manufacturing industries and capitalist economies developed, and cities grew to absorb more than half the world's burgeoning population. We created the Anthropocene. Science increased exponentially, differentiated into a kaleidoscope of disciplines and specialties, became a profession, came to rely upon state research funding, and acquired a diverse and demanding constituency.

In 1918, a watershed moment in the transformation, Max Weber spoke of science as a vocation to an audience of students at the University of Munich. He likened a calling or vocation for science to a life in the clergy: A person called would experience "a strange intoxication, ridiculed by every outsider" and would feel that "the fate of his soul depends upon whether or not he makes the correct conjecture at this passage of this manuscript."[8] Notably, Weber darkly described the "state capitalist ... institutes of medicine or natural science.... [wherein] the assistant's position is just as precarious as any 'quasi-proletarian' existence."[9] Seventeenth-century science was a wealthy gentleman's avocation, but by the early twentieth century the scientific life was on the cusp of turning from a vocation heard deep within the soul to a "quasi-proletarian existence" of alienation, dependence, and autonomy lost.[10] All in all, this is hardly the life of the lynx imagined by Cesi.

Science since Weber has continued to change, and changing circumstances impose new demands on the organization of science, the institutions (rules, laws, principles, and ethics) that guide it, and on its place in society. Within the environmental sciences, problems of food and water security, climate change, energy, urban redesign, demographic dynamics, disease vectors, and the coupled dynamics of natural and human systems pose unprecedented challenges to our ability to understand and act.[11]

What is the bridge from the utopian visions of the seventeenth century to the present? It is the enduring quest to create novel, generative organizations and institutions that embody earlier visions of science as intrinsically motivating—a vocation—and as a force for human betterment. Doing so will require rethinking how research is organized and conducted, and utopian visions are an inspiring though fragile platform for the task. To narrow the gap we draw upon original research on two environmental research organizations which, though quite different, have been exceptionally successful at producing highly creative scientific collaborations capable of contributing to human well-being by enabling a generative form of scientific interaction which we term "intellectual fusion." We do so by using ideas from social theory, forward-looking programmatic statements,[12] and principles derived from studies of small-group creativity that suggest design elements that might, in combination with complementary policies and strategies, work to promote the interdisciplinary, innovative, transformative research needed today.[13]

3.3 New Organizational Form and Function

The origin, design, and operation of two recently formed research organizations in the environmental sciences bear the architectural signature of a new Salomon's House. This section describes each organization, sketching its origins and organizational structure.[14] A methodological appendix explains our research approaches and outlines the data gathered about each organization.

3.3.1 National Center for Ecological Analysis and Synthesis

The National Center for Ecological Analysis and Synthesis (NCEAS) was founded in May 1995 through a cooperative agreement between the National Science Foundation (NSF) and the University of California. A national center supported by public funds at a level of about $4 M–$5 M per year, NCEAS serves the entire community of ecologists and environmental scientists. It has a director, associate director, and professional staff; a science advisory board; and reporting relationships both to the NSF and to an academic unit within the University of California, Santa Barbara. Roughly every three years it is evaluated by an outside panel of scientists.

A center of this magnitude is not the work of a single mind or moment but is instead a collaborative response to changes in the environment of ecological sciences. This response occurred on several levels and involved extensive discussions among funding agency officials, representatives of scientific societies, and scientists about the imagined center's rationale, mission, and design. O.J. Reichman, a program officer at the NSF, summarized a year or more of background discussions by observing that "ecological research problems are inherently multidisciplinary, requiring the efforts of biologists, engineers, social scientists and policymakers for their solution. Hence, there is a need for sites where a longer-term, multidisciplinary analysis of environmental problems can be undertaken."[15] About a year later, the Ecological Society of America and the Association of Ecosystem Research Centers convened a workshop to outline the "scientific objectives, structure, and implementation" of a "National Center for Ecological Synthesis." Their joint report concluded that "Knowledge of ecological systems is growing at an accelerating rate. Progress is lagging in *synthetic research* to consolidate this knowledge base into general patterns and principles that advance the science and are useful for environmental decision making. ... Without such *synthetic studies*, it will be impossible for ecology to become the predictive science required by current and future environmental problems."[16]

NCEAS gave form to an emergent understanding among ecologists that ecological research was becoming more collaborative, interdisciplinary, and engaged with policy, practice, and resource management; the scale of analysis was extending in time and space from contemporary studies of small sites to longitudinal studies integrating data across widely dispersed sites to examine broader temporal and spatial processes; the analytic techniques increasingly involved mathematical models estimated by increasingly sophisticated computers integrating secondary data from globally distributed field sites.

To meet these challenges, NCEAS developed a distinctive mode of collaboration: temporary working groups convened to engage in deep analysis and synthesis of existing theory, data, and methods about a specific scientific topic or policy issue. Groups typically consist of 8–15 collaborators who convene at NCEAS for about a week, several times each year, over two or three years. These intense periods of face-to-face collaboration are complemented by "homework" and electronic collaboration conducted by group members during the intervals between meetings. Unlike the academic "families" of advisors and graduate students of traditional ecology, NCEAS working groups are larger, more diverse collaborations spanning disciplines and extending from academe into the worlds of policy and practice.

3.3.2   Resilience Alliance

The Resilience Alliance (RA) has its intellectual origins in C.S. Holling's (1973) "Resilience and the Stability of Ecological Systems," the theoretical cornerstone for resilience work and a classic in ecology.[17] Overlooked for about 20 years, interest in the concept of resilience grew in the early 1990s when the Beijer Institute for Ecological Economics incorporated resilience theory into its workshops on biodiversity. In 1996, Holling and colleagues secured funds from the MacArthur Foundation and the University of Florida to develop an international, interdisciplinary research network focusing on resilience research. Members collaborated during a series of weeklong workshops, organized at nine-month intervals on remote islands in various locations worldwide. RA was formalized in 1998, with support from the Rockefeller and McDonnell Foundations.

Unlike NCEAS, RA is not a formal research organization but rather a loosely organized network of like-minded scientists and environmental managers and practitioners collaborating in ad hoc groups to advance ecosystem science and management. RA began with a core set of seven scientists, took on 23 new scientists (8 junior, 15 senior) over the next ten years, and added 20 scientists to their ranks since 2006. At this writing, approximately 40–50 scientists claim a close RA affiliation. Members are experts in ecosystem ecology and management, applied mathematics, natural resource management, social vulnerability studies, ecological economics, and political science. The group has admitted proportionately more social scientists over time, and some original members have retired or moved on to new lines of research. Beyond these are hundreds of researchers who have published resilience research or attended the large-scale international resilience conferences without deeper connections to the RA.

A true network organization, RA does not occupy a particular building or place but comprises 17 member "nodes" at universities and international research centers. Each node pays an annual membership fee to support RA's operations and journal, organizes local resilience research, and recruits members. Node representatives attend RA "full synthesis" meetings "on island" 1998 approximately every 18 months, where they share local contributions to resilience science, synthesize findings and theory, and decide on future directions for the organization. RA expects to live its theory by undergoing a process of decline and transformation, and so its future rests in the hands of the RAYS (Resilience Alliance Young Scholars) and their intellectual offspring, and will likely be transmitted as a cultural propensity to think and work in distinctive ways, rather than as an intact and functioning network.

3.3.3   Distinctive Forms of Knowledge

NCEAS and RA were founded around the same time by quite different means. NCEAS arose from within the scientific establishment, as represented by the NSF and a coalition of scientific societies. While there was vocal opposition to NCEAS—many feared its cost would reduce funds available for individual-investigator grants in ecology (a topic discussed by Feller, this volume)—it was designed and endorsed by the major professional associations of US ecology and has become so well established that many cannot imagine how the field will manage without the center.[18]

The RA, in contrast, arose in opposition to established ecological theory and grew incrementally as the network expanded, supported by funding from foundations. Differences aside, the organizations are similar in their orientation to a grand idea (transforming ecology, resilience), novel collaborative patterns and processes, and engagement with real-world problems in ways that also advance fundamental knowledge (use-inspired fundamental research).[19]

Scientific synthesis is the integration of disparate theories, methods, and data across disciplines, specialties, professional sectors, and spatial or temporal scales to produce models and explanations of greater generality, parsimony, or completeness.[20] Synthesis is vital for a future in which increasingly specialized sciences and professions face integrative intellectual questions and pressing problems that demand coherent application of ideas drawn from diverse fields of expertise. Synthetic explanations exhibit emergent properties that differ from their constituent elements, and explain phenomena that span disciplines or extend across spatial or temporal scales.[21] RA and NCEAS are among the first organizations to use the term "synthesis" in this sense to characterize their work. The idea has spread rapidly across disciplinary and national borders, and has drawn substantial policy attention and investment: some 18 synthesis centers in various fields are at work today.

The intellectual impetus for NCEAS, which continues shaping the conduct and content of its research, is the call for "ecologists to look outward rather than inward to integrate extensive information across disciplines, scales, and systems."[22] Research performed at NCEAS differs from the traditional field-based science of ecology. Most ecological studies are conducted in small areas (a few square meters) for short amounts of time (a field season) by disciplinary groups (that resemble families, with a senior scientist accompanied by intellectual offspring), while the focus at NCEAS is larger in scale, longer in time, interdisciplinary, and often applied in orientation. Where traditional ecology involves prolonged, hands-on fieldwork, NCEAS scientists are seldom familiar with the study sites from which their data were gathered, relying instead on metadata to render data useable; trips to the field are replaced with advanced statistical analysis and mathematical modeling. Finally, traditional ecological research tends to reduce general hypotheses to empirical tests conducted in particular places, whereas NCEAS research specifically seeks to uncover general laws, emergent properties, and broadly applicable theories and management solutions.

NCEAS working groups collaborate off-campus and blend proximate, face-to-face work with distal, computer-mediated interaction. Potential uses and users of research are an intrinsic part of the research process, and groups and their intended audiences transcend disciplines and the usual bounds of academic collaboration. Some 3400 scientists have taken part in NCEAS activities, representing 49 countries, 531 different academic organizations, 428 non-academic organizations (such as government agencies, companies, and non-governmental organizations), and more than 360 scientific societies. More than a quarter of NCEAS working groups address issues of environmental policy, resource management, conservation or other applications. Practical research aims, such as creating a marine reserve or designing a fisheries management plan, are entwined with the academic aims of scholarly publication. At present NCEAS has produced over 1000 publications, including 41 in *Science*, 26 in *Nature*, and 21 in the *Proceedings of the National Academy of Sciences*. It is among the top 1% most cited ecological research organizations.

Resilience theory similarly challenges and destabilizes established ecological theory and resource management science. RA members coined the term resilience in ecology and led the development of resilience theory, a systems perspective that blends ideas from complexity theory, ecosystems ecology, and social sciences to understand dynamics of transformation in social–ecological systems.[23] Traditional ecological theory assumes that ecosystems are closed, self-regulating systems distributed around equilibrium states. Human activities are viewed as "disturbing" these "natural" processes, and so are downplayed or excluded. In contrast, resilience theory seeks to integrate natural and social systems, emphasizes non-linear dynamics, multiple stability domains, dynamic and stochastic change, and replaces economic optimization strategies with dynamic stochastic models.[24] Disturbance by humans is viewed as ubiquitous and normal, rather than uncommon, rendering humans licit subjects of ecological analysis.

RA research has substantially influenced environmental science and policy. Members regularly publish in top journals and receive major academic awards, and RA research has informed natural resource management in the USA, Europe, Australia, South Africa, and other countries. Its ideas were discussed during the World Development Summit, incorporated into the Intergovernmental Panel on Climate Change, and are used by The World Bank and other organizations. Members have founded major international research centers, including the Stockholm Resilience Center and the South American Institute for Resilience and Sustainability Studies. Over 600 participants attended the first international conference on resilience in Stockholm (2008), and more than 800 attended the second in Tempe, AZ (March 2011). How is this accomplished? The deeply original knowledge produced by NCEAS and RA is made possible by distinctive patterns and processes of collaboration that enable intellectual fusion.

3.4   Group Creativity and Intellectual Fusion

Collaboration, in science and as in other creative endeavors, can give rise to a climate conducive to ideas and insights beyond the powers of those present. Both creative process and its accomplishments have a quality that is seldom experienced but immediately recognized. In the words of Ludwik Fleck:

> He is a poor observer who does not notice that a stimulating conversation between two persons soon creates a condition in which each utters thoughts that he would not have been able to produce either by himself or in different company. A special mood arises, which would not otherwise affect

either partner of the conversation but almost always returns whenever these persons meet again.[25]

Scientific synthesis, or the integration of concepts, theories, and data into original and potentially transformative explanations, results from creative processes within groups. We will offer a synthesis of theories of group creativity and propose a new idea, intellectual fusion, to describe particular processes that promote the integration of disparate concepts, theories, and data. The concept of intellectual fusion draws upon and contributes to two streams of theory. The first defines creativity as the novel recombination of dissimilar components into useful new patterns.[26] The second identifies characteristics of group organization and dynamics that promote creativity.[27] Unlike trading-zone theories of interdisciplinary collaboration, where the exchange of "finished" intellectual goods is facilitated by interactional expertise,[28] theories of group creativity are concerned with the organizational patterns and micro-social processes that combine ideas and evidence to form strikingly original explanations. In doing so, this line of theorizing and research addresses a challenge posed by Daniel Stokols and colleagues in the *Oxford Handbook of Interdisciplinarity*:

> ...the scientific outcomes of TS [team science] initiatives are strongly influenced by social and interpersonal processes, including team members' collaborative styles and behaviors, interpersonal skills, and negotiating strategies. Yet the precise ways in which these social processes—such as team members' disagreements about scientific issues, interpersonal trust, 'group think' among scientists who had worked together over extended periods—influence scientific productivity and TD [transdisciplinary] integration are not known.[29]

Theories of group creativity identify a remarkably similar set of causal factors and conditions, although their elements have different names and their mechanisms differ or remain unclear. Randall Collins proposes that creative intellectual work results from emotional energy combined with cultural capital.[30] Ellen Jane and J. Rogers Hollingsworth recast this insight into a form compatible with organizational theory, proposing that the level and diversity of expertise (capital), in conjunction with an organizational design that promotes dense and enduring interactions (energy), explains why particular universities have been home to major discoveries in the biomedical sciences, while others (of equal distinction and greater size) were not.[31] Over several decades' work Teresa Amabile has developed a "componential theory" of creativity that has four major parts: skills, environment, motivation, and creative processes.[32] According to Amabile, creative individuals and groups require particular conditions in each of these four categories: skills, environment, motivation, and processes (such as curiosity and risk-taking). Our work builds upon these ideas by developing the concept of emotional energy and outlining its role in science, and by generalizing the concepts of skill and cultural capital to various forms of capital (technology, human, social, cultural). We also develop and illustrate the concept of "intellectual fusion" to describe at a finer-grained level the creative integration of ideas and evidence.[33]

By analogy with the process of nuclear fusion, we propose that distinctive socio-emotional processes occurring within collaborative groups disassociate concepts, methods, and theories from their disciplines, paradigms, and professions of origin, recombining them to form original and useful configurations. We further propose that these group processes account, at the micro-social level, for the differences between universities that Hollingsworth reports (at the organizational level), and that fusion and its antecedent conditions result from particular forms of social organization combined with specific environmental and cultural conditions. We will first outline nuclear fusion and its intellectual analog, then discuss and illustrate the conditions that encourage it, which we call resources (talent, in Amabile's theory of individual creativity), context (environment, to Amabile), emotional energy (motivation), and alternation (creative dynamics or processes).

Nuclear fusion occurs when two or more atomic nuclei fuse to form a single, heavier nucleus, releasing energy which, under appropriate conditions, sustains the reaction. Fusion depends first of all on the presence of the right elements in the right proportions. Then energy is added to form a plasma, which must become energetic (hot) enough, dense enough, and endure long enough for nuclear collisions to occur. Collisions are rare, and when they do occur they must be energetic enough to overcome the electrostatic force at the surface that acts at a distance to hold nuclei apart, and reach the point where the stronger, attractive nuclear force overcomes resistance and causes them to fuse. Fusion not only forms new elements but also releases energy that sustains the reaction. Energetic plasmas are active, almost alive, and highly reactive: if one contacts the walls of the containment vessel it becomes contaminated and cools, extinguishing the reaction, and so the plasma must be isolated from its surroundings.

Think of scientific disciplines as assemblies of theories, methods, research technologies, techniques, orienting research questions, standards of evidence and proof, and such bound together to form a more or less coherent disciplinary matrix[34] or epistemic culture.[35] That training, and the particular ontologies, epistemic cultures, and research systems it imparts, creates the conditions that cause the mutual incomprehension that separates disciplines, promotes resistance to collaboration, and impedes synthesis (see editors' introduction on the late writings of Kuhn).[36] Stated simply, the highly diverse collaborations of scientific synthesis are fundamentally unnatural acts, eliciting strong resistance that can only be overcome through the creation of specific social and environmental conditions. Intellectual fusion occurs when the appropriate conditions (context, resources, energy, and dynamics) are present in proper proportions and amounts to overcome resistance to inter-sectoral or interdisciplinary collaboration imparted by training and maintained by intellectual and organizational sanctions. When fusion occurs, propositions, concepts, or ideas held together in the disciplinary matrix become less strongly associated with one another and available to fuse into novel combinations (the process resembles a lowering of critical or skeptical inhibitions).[37] As indicated by the theories above, intellectual fusion requires a combination of *context* (e.g., isolated from distraction and steeped in mutual trust), *resources* (intellectual ability, social capital, data, and research technology of diverse types), *emotional energy* (to ignite and sustain interaction), and group *dynamics* that alternate appropriately between competing values. We will discuss each below.

3.4.1 Context

An organizational *context* that is isolated or well insulated from surrounding distractions is essential for overcoming resistance and achieving fusion. The most striking feature of a nuclear fusion reactor is the magnetic "bottle" that contains the plasma and isolates it from the environment (including from the vessel itself), because if the plasma contacts the vessel walls it will become contaminated and cool, ending the fusion reaction. In similar fashion, a collaborative dynamic cools and loses energy and focus when its members become distracted by other purposes (email, texts, or visits with local colleagues) in the course of a collaborative work session.

For RA an experience that they call "island time" creates the isolation and other conditions necessary for fusion to occur. Islands are isolated, neutral locations that limit distraction and enhance focus.[38] NCEAS is located in downtown Santa Barbara, ten miles from campus, to discourage working group participants from wandering off to visit friends. Extended, exclusive contact with a small group of collaborators working in physical and social isolation lowers resistance to interdisciplinary collaborations. For example, the core ideas of *Panarchy*, the canonical resilience treatise, "were developed, tested, and modified in a series of workshops (...) held on an 'island'—where we were in a sense isolated from the outside world and free to explore, argue, contrast, and test the concepts that are in this volume."[39]

Interpersonal trust is foundational for the functioning of science[40] and essential for success in interdependent, big-science collaborations.[41] A degree of "instrumental intimacy" increases the density of interactions by reducing participants' wariness or self-censorship, creating conditions in which "partners will share their most half-baked ideas, trusting that others will not destructively attack or plagiarize them."[42] The necessity of trust is heightened early in a research endeavor, where "free, unmonitored exchanges about unpublished results and ideas require powerful norms to protect the individual's priority of discovery."[43]

Deep trust in various senses is vital for groups that challenge traditional disciplines: trust in others' substantive expertise and abilities, trust that they will keep confidences and respect others' ideas (including both evaluating them fairly and not stealing them), and trust that they will do as they promise when island time ends. An RA member said:

> Each of us trusts our colleagues to do what they purport to do, that indeed they are truly [involved] in an honest effort of mutual discovery and have fun in mutual discovery. Their words are trusted, their discoveries are trusted, their ideas are trusted. Trust is an important element and it's based upon the assumption that that trust and imagination that individuals have [on island] can persist [off island] when most of the forces on the individual are local and institutional.

Trust enables intense collaboration in isolated locales, bold (but perhaps not entirely sound) conjecture, rapid and frank criticism (peer review), and intellectual fusion. Too much trust, particularly when combined with isolation and group solidarity, can lead to a closed-minded, uncritical acceptance of ideas from within ("groupthink").[44] Effective groups alternate between the competing values of constructive versus critical modalities.

3.4.2 Resources

Several forms of *productive resources* (or capital) are essential for intellectual fusion: the scientific and social intelligence, education, and skills of people involved in the collaboration; the extent and quality of connections that reach from the collaboration into the wider world; the research technologies of data, instruments, analytic tools, and computational resources available to produce knowledge.[45] For intense, isolated, and episodic collaborations to succeed, participants must bring the necessary resources or be able to reach them rapidly (e.g. by phone or internet) because time is short and work proceeds rapidly. Experts must also be willing to make their expertise available to others—a notable concern in the increasingly competitive and proprietary culture of contemporary science—and the recipients or beneficiaries of expertise must have sufficient "interactional expertise" to access and use the knowledge imparted.[46] In this dimension, the RA is more isolated and autarkic; NCEAS is more dependent on interpersonal and computational networks.

As a strict rule, NCEAS and RA organizers invite only scholars with excellent scientific ability *and* "good island personalities": the sort of people who have deep and extensive knowledge, are willing to share it, and are able to elicit and make use of others' expertise.[47]

Another resource is the diversity of expertise, social and cultural background, employment sector, values and ethical commitments, or area of responsibility available for synthesis (fusion), because these are the raw materials that are recombined into something original and useful. Studies of small, task-oriented groups show that diversity, within bounds, enhances performance and creativity. Too much diversity (particularly social diversity, rather than technical or task-relevant diversity) increases centrifugal forces that will pull a group apart; too little and there is not enough dissimilarity to fuel originality.[48] For group diversity to matter for performance it must also be equitably deployed: each member of the group must be afforded opportunities to contribute what he or she can to the collective enterprise (a sort of social stoichiometry).[49] Merely having diverse abilities and perspectives in the room is not sufficient for the group to benefit: "group intelligence" is activated by the equitable (not necessarily equal) participation of all members. One group leader only invited scientists who "don't have big egos...we've had people invited to these meeting and they did have egos. And they don't get invited back, no matter how clever they are" because big egos are repulsive forces that interfere with the open exchange of ideas by dominating the conversation or crippling others with criticism.

3.4.3   Energy

Intense face-to-face intellectual exchanges produce high levels of *emotional energy*, a force that drives intellectual fusion by instilling enthusiasm and commitment in a collaborative group.[50] NCEAS working groups generate emotional energy during occasional but intense face-to-face interactions that last from several days to a week. Scientists reside in the same small hotel, breakfast and walk together (two kilometers) to the Center, and spend evenings having drinks and dinner in local restaurants. Informal social interactions and rituals that extend beyond the working day generate and sustain emotional energy and group solidarity. By positioning themselves in opposition to dominant intellectual trends, a group's ideas, writings, and thought style awaken feelings of solidarity among members and separation from others. They also imbue intellectual grievances with deep emotional significance,[51] and so "Words which were formerly simple terms become slogans; sentences which once were simple statements become calls to battle.... They no longer influence the mind through their logical meaning."[52] Dense and enduring interactions combine to increase the chances that useful ideas will engage one another: the longer and more intense the interactions within a group, and the more focused they are, the more likely it is that complementary ideas will collide and bond.

3.4.4   Group Dynamics: Ambivalence and Alternation

As discussed by the editors of this volume in the introduction, for several decades scholars have recognized that the culture of science is characterized by ambivalence or contradiction, and that values in tension are essential to—even constitutive of—science.[53] Thomas Kuhn detected an essential tension between tradition and originality at the heart of science, and inferred that "[t]he ability to support a tension that can occasionally become almost unbearable is one of the prime requisites for the very best sort of scientific research."[54] Kuhn's insight gained empirical support in the path-breaking organizational research of Donald Pelz and Frank Andrews, who concluded that "[t]he optimum climate [for high-quality scientific research] was not necessarily some compromise between extremes. Rather, achievement often flourished in the presence of factors that seemed antithetical."[55] Ambivalence structures the organization and working arrangements of science: a set of polar conditions or values are in tension, each pole in itself is desirable, but the highest levels of performance depend upon the activation of the one value that best suits a particular circumstance (much as some genes control the expression of others). Along these lines Keith Sawyer observes that "group flow happens when many tensions are in perfect balance: the tension between convention and novelty; between structure and improvisation; between the critical, analytical mind and the freewheeling, outside-the-box mind; between listening to the rest of the group and speaking out in individual voices."[56]

We propose, however, that "perfect balance" is not "a compromise between extremes" or a golden mean but is instead a dynamic equilibrium achieved through the *alternation* between contradictory values, principles of organization or conditions of work, *activating* first one, then the other, as conditions and circumstances warrant.[57] The balance will be dynamic and so it will not be perfect; instead, the appropriate proportions of one quality or another, the appropriate shift from one pole to another, will be determined by contingencies of organization, interaction, psychology, and the substance of the work itself in ways that remain to be explored. That exploration would first list the values in tension, then identify mechanisms that switch a group from one value pole to another, and conclude by demonstrating that successful alternation matters for performance.

In our studies of the synthesis centers three value pairs that are intimately involved in scientific practice and performance have been managed by alternation. The list is incomplete and illustrative; additional tensions are discussed elsewhere.[58]

Constructive—Critical While it is overly simple to say that science oscillates between contexts of discovery and justification,[59] conjecture and refutation,[60] or originality and tradition,[61] anyone who has observed scientists at work for any length of time has noticed that they build explanations (theories, results, conclusions) for a time, and then test, challenge, or criticize them. The intervals may be longer or briefer, the roles of builder and critic may be specialized or rotate, and the exercises of skepticism may be gentler or rougher, but activities of both sorts are part of the research process. Suspending criticism or disbelief, at times nearly to the point of partisanship, is an essential element of creativity. Mitroff's (1974) study of the Apollo moon scientists found that the most creative scientists were also the most emotionally committed to their theories. Millikan evaluated his oil-drop experiment data with powerful preconceptions about the correct theoretical frame and empirical outcome.[62] The history of science is littered with ideas and findings that were initially rejected or ignored but later rediscovered and accepted[63]: unsubstantiated commitment to an idea or perspective may be the price of saving potentially transformative ideas from early demise at the hands of skeptics.

Much of the RA's on-island collaboration is done with criticism (disbelief) suspended: people finish one another's sentences, think one another's thoughts, and seldom is heard a discouraging word (with the near-lethal exception of the Malta Affair, which brought skeptical, feisty, disciplinary minds into the mix).[64] RA members are aware that they have created this protected space:

> The culture of science is dominantly skepticism, and appropriately so. But that is not true in [RA] .... Rather, the culture is much more focused on the generation of innovative ideas and testing. But not skepticism, so it is a very, very different culture.

For RA criticism happens later, off-island, when work is prepared for publication and tested by peer review.

NCEAS working groups alternate much more rapidly between constructive and critical modalities, with peer review accomplished on the fly, interleaved with speculative ideas and explanations. Key to their research process is the immediate, free-form, and interactive challenge and riposte that anneals a novel idea in the fires of skepticism. We observed a group that had more than 50 critical exchanges—expressions of skepticism or evaluation and reaffirmations of the original claim—within the space of an hour, with rising levels of emotional heat.[65] Group cohesiveness and trust (contextual elements) are essential for a group to withstand such intense critical exchange. In return, the reward for doing so is increased velocity of research that sorts through ideas, data, and literature rapidly and with strong purpose and critical acumen.

Criticism must enter at some point or a group is at risk of self-deception or "groupthink."[66] The phage group established a practice of devastating critiques of one another's work, which improved quality, instilled boldness (because criticism created a safety net that trapped bad ideas), and tempered commitment. No outside criticism could equal that which was inflicted from within.[67] This was enabled by the intense social bonds that held the group together, just as strong social bonds and trust facilitate concise and pointed criticism—peer review on the fly—within RA or an NCEAS working group.

Universal–Local This is the tension between the quest for generalizable or universal laws and theories versus the pursuit of context-specific explanations and interventions tailored to a particular place, time, challenge, or purpose. Farmers, for example, mistrust science-based advice unless they see it demonstrated on their land[68] because subtle differences in elevation, orientation, contour, soil composition, permeability, and more are local contingencies that challenge the credibility, salience, and legitimacy of general knowledge applied to a specific place.[69] A similar integration of the general and the particular occurs in efforts to enhance the visibility of consequences[70] or inspire fundamental research with the needs and possibilities of particular uses.[71]

NCEAS and RA commute between the worlds of knowledge and application. For example, a long-running and highly productive NCEAS group was concerned with the design, operation, and effects of marine protected areas. By focusing on the requirements of a *specific* marine reserve in a *particular* patch of ocean, while also pursuing answers to basic scientific questions (they published 36 articles), a collaboration among disciplines and sectors (academe, government, fishermen, NGOs) produced fundamental research papers and design principles for a working reserve.

Dirigism–Autonomy Universalism and organized skepticism, two of the cardinal values in Merton's institutional model of science, imply a democratic leveling among scientists, a community of equals separated only by their ability to contribute to the production of certified knowledge. A scientist interviewed by Warren Hagstrom in the early 1960s summarized this view in these words: "Telling someone what to do is *taboo*. The greatest man in science cannot tell the lowest what to do."[72] Scientists interviewed decades later offered similar views, expressing something approaching a right for fully fledged (i.e. doctoral) scientists to determine the course of their research.[73] And something near equality, extended to graduate students, altered the outcome of a working group at NCEAS (discussed above).

Yet there is an equally long established countervailing inclination toward imperative coordination or directedness and away from democratic and deliberative decision making in science. For example, in his autobiography François Jacob characterized Boris Ephrussi, "without doubt the most outstanding figure in French biology" of his time, as "very domineering. Ephrussi ruled his laboratory and his students with an iron hand. He did not hesitate to throw down the sink an experiment one of his students had taken the liberty of performing without asking his opinion."[74] Decades later a young laboratory head at a private US university would first profess that "Everything that happens in the lab is the consequence of a discussion between me and the postdocs," then continue by asserting quite the opposite: "I don't have a problem being autocratic .... I am in charge of this lab. It has to be that way in science. You can't do science in a democratic way, because it has to be one way of thinking. Maybe the wrong way of thinking, but it has to be one way."[75] Democratic and directed modalities of decision making coexist within the culture and practice of laboratory leadership, and each has demonstrated effectiveness in various contexts.[76]

3.4.5   Group Flow

When context, resources, energy, and dynamics are optimal, groups experience a phenomenon called "flow": a state of heightened consciousness, sharpened attention, and total immersion in the task at hand, which is accompanied by diminished self-consciousness, distorted perceptions of time, and a feeling of personal control over events.[77] This may be the intersubjective sensation of intellectual fusion at its fullest. Flow is experienced by jazz ensembles, basketball teams, and other small groups.[78] RA members experienced collaborative flow, which facilitates highly focused discussion wherein "new ideas seem to emerge from the dialogue without 'belonging' to [anyone], and afterwards they may not be able to say who had the ideas first."[79] Things happen so quickly that time seems to slow, increasing the velocity and efficiency of collaboration:

you can almost communicate by minimal sounds. It's almost like you know people so well that one of them makes a head gesture like that [nodding] and it just communicates a whole subroutine of knowledge at once... an enormous volume of information per unit time gets transmitted.

The confluence of these conditions produces a powerful experience, as a senior RA member recalled:

we became like more or less a collective brain, or a collective soul. We were sitting out on the porch—about five or six of us.... And we started to talk, and then suddenly after a while you couldn't any longer feel who talked about what, it was like a unified experience...—just that 45 minutes or so— it was sort of, I wouldn't say another level of consciousness, but level of communication that generated new insights. And you couldn't really say afterwards who had said what. That was fantastic.

Note that the language—vision, intuition, soul, artistry—is not the typical analytic, dispassionate vocabulary of scientists.

3.5 SALOMON'S HOUSE REIMAGINED

In this chapter we have explored how one particular innovative organizational arrangement—synthesis centers and the collaborative working groups they catalyze and host—generate, develop, and institutionalize the path-breaking intellectual contributions that spark the continual renewal of science. Our contribution complements recent work on interdisciplinarity[80] and typologies of trading zones[81] but is quite distinct from that work. Our specific contribution links organizational form and formation through group patterns and processes, understood at the micro-social level, to creative and impactful scientific and practical outcomes. Drawing upon theories of creativity from psychology and sociology, we outline a general model of conditions that promote group creativity (context, resources, energy, and dynamics), show its relevance for synthesis centers, and describe a particular mechanism—intellectual fusion—that occurs under such conditions to meld diverse ideas and evidence into original and useful results. The forces of renewal that have led to synthesis and synthesis centers continue unabated, and so it is worth asking where this is leading: what further forms of organizational and institutional innovation are on the horizon?

One incipient innovation departs from the current image of dispassionate, disengaged scientific inquiry conducted without considering possible uses and their implications. This model has been in retreat for nearly 20 years, replaced by a form of transdisciplinary science that has *visible practical consequences*, strives to improve well-being, and takes on urgent challenges. Disciplinary science will be complemented—not supplanted—by heterogeneous or transdisciplinary collaborations that are born in common cause and energized by commitment and emotion. Such collaborations will be difficult to manage, as incompatible epistemic commitments will exist alongside competing economic and political interests.

Science of this sort will necessarily have explicit value engagements and commitments, and these must be addressed openly rather than excluded by assumption (which only allows them to return in less tractable form: see "climategate," helicobacter pylori,[82] the Apollo moon scientists,[83] plate tectonics,[84] even the Milliken oil-drop experiment,[85] among many examples). What matters is not that scientists have strong value positions: of course they do, and they must, in order to work with energy and intensity.[86]

Similarly, rather than denying the emotional dimension of science, which is present even in the pursuit of fundamental knowledge and intensified when research engages urgent, real-world problems, we will instead devise arrangements that generate and direct emotional energy (or "hot thought").[87] Value commitments and emotional energy will be tempered by a more robust peer review process that is continual, interactive, iterative, and diverse in who counts as a peer, including relevant professions and members of the public. Their review would be a comprehensive dialogue that evaluates reasoning, evidence, inferences, and implications with a recurrence and intensity that can be sustained only when embedded in a matrix of mutual trust.

In the Nicomachean Ethics (c. 350 BCE), Aristotle placed episteme and techne—science and technology—atop his list of intellectual virtues, and put at their side the virtue of phronesis, which is "practical wisdom" or the situational ethics that tell us the right thing to do in particular circumstances.[88] Aristotle understood that for science and technology to serve their social purposes they must be guided at an intimate, working level by ethical precepts and reasoned value positions. Phronesis now must be cultivated as both a personal virtue and as an organized, collective capability to form, in partnership with episteme, and techne, a sturdy tripod for the design of research organizations and the conduct of research.

NOTES

1. Francis Bacon, *The New Atlantis* (London, 1627).
2. Peter Weingart, "A Short History of Knowledge Formations," in *The Handbook of Interdisciplinarity*, ed. Robert P. Frodeman, Julie Thompson Klein, and Carl Micham (New York: Oxford, 2010), 14.
3. Bacon, *The New Atlantis*, 19.
4. Bacon, *The New Atlantis*, 24.
5. Judah Bierman, "Science and Society in the New Atlantis and Other Renaissance Utopias," *Proceedings of the Modern Language Association* 78 (1963): 499–500.
6. Johann Valentin Andreae, *Christianopolis: An Idea State of the Seventeenth Century* (New York: Oxford University Press, 1916[1619]).
7. From the Lynceographia [1612], quoted by Margaret Ornstein, *The Role of Scientific Societies in the Seventeenth Century* (Hamden, CT: Archon Books, 1963), 75.
8. Max Weber, "Science as a Vocation," in *From Max Weber: Essays in Sociology*, ed. Hans H. Gerth and C. Wright Mills (New York: Oxford, 1946 [1918]), 135.
9. Weber, "Science as a Vocation," 131.
10. Edward J. Hackett, "Science as a Vocation in the 1990s: The Changing Organizational Culture of Academic Science," *Journal of Higher Education* 61 (1990): 241–279.
11. Johan Rockström et al., "A Safe Operating Space for Humanity," Nature 461 (2009): 472–474; Ruth DeFries et al., "Planetary Opportunities: A Social Contract for Global Change Science to Contribute to a Sustainable Future," *BioScience* 62 (2012): 603–606.
12. Stephen Carpenter et al., "Accelerate Synthesis in Ecology and Environmental Sciences," *BioScience* 59 (2009): 699–701; STEPS Centre, *Innovation, Sustainability, Development: A New Manifesto* (STEPS: Sussex, England, 2010).
13. Principally, the ideas we use are drawn from the theory of scientific and intellectual social movements: Nicholas C. Mullins, *Theories and Theory Groups in Contemporary American Sociology* (New York: Harper, 1973); Scott Frickel and Neil Gross, "A General Theory of Scientific/Intellectual Movements." *American Sociological Review* 70 (2005): 204–232; John N. Parker and Edward J. Hackett, "Hot Spots and Hot Moments in Scientific Collaboration and Social Movements," *American Sociological Review* 77 (2012): 21–44; the theory of interaction ritual chains: Randall Collins, *The Sociology of Philosophies* (Cambridge, MA: Harvard University Press, 1998); and research on collaborative circles: Michael P. Farrell, *Collaborative Circles: Friendship Dynamics and Creative Work* (Chicago:

University of Chicago Press, 2001); Ugo Corte, *Subcultures and Small Groups: A Social Movement Theory Approach* (Uppsala: Dissertation, Uppsala University, 2012).

14. More information about the cases is available in Edward J. Hackett et al., "Ecology Transformed: The National Center for Ecological Analysis and Synthesis and the Changing Patterns of Ecological Research," in *Scientific Collaboration on the Internet*, ed. Gary Olson, Nathan Bos, and Ann Zimmerman (Cambridge, MA: MIT Press, 2008); Stephanie H. Hampton and John N. Parker, "Collaboration and Productivity in Scientific Synthesis" *Bioscience* 61(2011): 900–910; Parker, Hackett, "Hot Spots"; John N. Parker and Edward J. Hackett, "The Sociology of Science and Emotions," in *Handbook of the Sociology of Emotions: Volume II*, ed. Jan E. Stets and Jonathan H. Turner (New York: Springer, 2014), 549–572; Edward J. Hackett and John N. Parker, "Ecological Science Reconfigured: Group and Organizational Dynamics in Scientific Change," in *The Local Configuration of New Research Fields: On Regional and National Diversity*, ed. Martina Merz and Philippe Sormani 153–171. New York: Springer.

15. Hans Reichenbach, *Experience and Prediction: An Analysis of the Foundations and Structure of Knowledge* (Chicago: University of Chicago Press, 1938), 1.

16. Ecological Society of America and the Association of Ecosystem Research Centers, *National Center for Ecological Synthesis: Scientific Objectives, Structure, and Implementation*, Report from a joint committee, based on a workshop held in Albuquerue, NM, 25–27 October 1992.

17. C.S. Holling, "Resilience and the Stability of Ecological Systems," *Annual Review of Ecology and Systematics* 4 (1973): 1–23.

18. NSF funding for NCEAS as the center is described here has ended, but the center has secured new funding sources from NSF (as the LTER Network Office) and from The Moore Foundation and The Nature Conservancy, along with new mandates to be even more applied and problem oriented. NSF is now funding, at twice the budget, the Socio-Environmental Synthesis Center (SESYNC) at the University of Maryland. SESYNC is modeled on NCEAS but has a broader mandate consistent with its larger budget. NSF also funds the National Evolutionary Synthesis Center (est. 2005) and the National Center for Mathematical and Biological Synthesis (est. 2008), further indicating its commitment to the scientific synthesis.

19. Donald Stokes, *Pasteur's Quadrant* (Washington, DC: Brookings, 1997).

20. Stephen Carpenter et al., "Accelerate Synthesis".

21. Brian Sidlauskas et al., "Linking Big: The Continuing Promise of Evolutionary Synthesis." *Evolution* 64 (2009): 871–880.

22. Ecological Society of America and the Association of Environmental Research Centers. National Center for Ecological Synthesis: Scientific

objectives, structure, and implementation. Report of a workshop held in Albuquerque, October 1992. (1993)

23. Resilience refers to the amount of change a system can absorb while maintaining its structure and function, its capacity for self-organization, and its capacity for learning and adaptation (see Parker and Hackett, "Hot Spots").

24. John Parker, "Integrating the Social into the Ecological: Organization and Research Group Challenges," in *Collaboration in the new life sciences*, ed. J.N. Parker, N. Vermeulen, and B. Penders (Burlington: Ashgate, 2010), 85–109.

25. Ludwik Fleck, *Genesis and Development of a Scientific Fact* (Chicago: University of Chicago Press, 1979 [1935]), 44.

26. Sarnoff A. Mednick, "Remote Associates Test," *Journal of Creative Behavior* 2 (1962):213–14; Edward M. Bowden and Mark Jung-Beeman, "Normative Data for 144 Compound Remote Associate Problems," *Behaviorial Research Methods, Instruments, & Computers* 35 (2003): 634–639; Diana R. Rhoten, Erin O'Connor, and Edward J. Hackett, "The Act of Collaborative Creation and the Art of Integrative Creativity: Originality, Disciplinarity, and Interdisciplinarity," *Thesis Eleven* 96 (2008): 83–108.

27. Teresa M. Amabile, "Social Psychology of Creativity: A Componential Conceptualization," *Journal of Personality and Social Psychology* 45 (1983): 997–1013; Teresa M. Amabile, "Componential Theory of Creativity," *Harvard Business School Working Paper* 12–096 (2012); Collins, *The Sociology of Philosophies;* Thomas Heinze et al., "Organizational and Institutional Influences on Creativity in Scientific Research," *Research Policy* 38 (2009): 610–623; J. Rogers Hollingsworth and Ellen Jane Hollingsworth, *Fostering Scientific Excellence: Organizations, Institutions, and Major Discoveries in Biomedical Science* (New York: Oxford University Press, 2012).

28. Harry M. Collins, Robert Evans, and Mike Gorman, "Trading Zones and Interactional Expertise," *Studies in History and Philosophy of Science* 38 (2007): 657–666.

29. Robert Frodeman, Julie Thomson Klein, and Carl Mitcham, *The Oxford Handbook of Interdisciplinarity* (Oxford: Oxford University Press, 2010), 477.

30. Collins, *The Sociology of Philosophies.*

31. Hollingsworth and Hollingsworth, *Fostering Scientific Excellence.*

32. Amabile, "Social Psychology of Creativity"; Amabile, "Componential Theory".

33. Rhoten, O'Connor, and Hackett, "The Act of Collaborative Creation"; Parker and Hackett, "Hot Spots"; Parker and Hackett, "The Sociology of Science".
34. Thomas S. Kuhn, *The Structure of Scientific Revolutions* (Chicago: University of Chicago Press, 1962).
35. Karin Knorr-Cetina, *Epistemic Cultures: How the Sciences Make Knowledge* (Cambridge, MA: Harvard University Press, 1999).
36. See editors' introduction on the late writings of Kuhn.
37. The concept of emotional energy also encompasses micro interpersonal social processes such as trust, "instrumental intimacy," and "escalating reciprocity." These are treated in fine detail by Farrell, *Collaborative Circles*.
38. To this end the phage group retreated to isolated locations such as the Anza Desert, and Bohr's quantum physicists went rock climbing: Belver C. Griffith and Nicholas C. Mullins, "Coherent Groups in Scientific Change: 'Invisible Colleges' May Be Consistent throughout Science," *Science* (1977): 959–64.
39. Lance Gunderson and C.S. Holling, *Panarchy: Understanding Transformations in Human and Natural Systems* (Washington, D.C.: Island Press, 2002), XXIII.
40. Steven Shapin, *The Scientific Life: A Moral History of a Late Modern Vocation* (Chicago: University of Chicago Press, 2008).
41. Knorr-Cetina, *Epistemic Cultures*.
42. Farrell, *Collaborative Circles*, 285.
43. Griffith and Mullins, "Coherent Groups," 962.
44. Irving L. Janis, *Victims of Group Think* (Boston: Houghton Mifflin Company, 1972).
45. This is a more varied and detailed set of resources than those included in Randall Collins' (1998) term "cultural capital," since we include material and social forms of capital, but they serve much the same purpose Collins, *The Sociology of Philosophies*; Mullins, *Theories and Theory Groups*; Farrell, *Collaborative Circles*.
46. H.M. Collins and Robert J. Evans, *Rethinking Expertise* (Chicago, IL: University of Chicago Press, 2007).
47. Farrell, *Collaborative Circles*.
48. Roger Guimerà et al., "Team Assembly Mechanisms Determine Collaboration Network Structure and Team Performance," *Science* 308 (2005): 697–702; John M. Levine and Richard L. Moreland, "Collaboration: The Social Context of Theory Development," *Personality and Social Psychology Review* 8 (2004): 164–172.
49. Anita W. Woolley et al., "Evidence for a Collective Intelligence Factor in the Performance of Human Groups," *Science* 330 (2010): 686–688.

50. Rhoten, O'Connor, and Hackett, "The Act of Collaborative Creation"; Parker and Hackett, "Hot Spots"; Parker and Hackett, "The Sociology of Science".

51. Rachel Schurman and William Munro, "Ideas, Thinkers, and Social Networks: The Process of Grievance Construction in the Anti-Genetic Engineering Movement," *Theory and Society* 35 (2006): 1–38.

52. Fleck, *Genesis*, 43.

53. Thomas S. Kuhn, *Essential Tensions: Selected Studies in Scientific Tradition and Change* (Chicago: University of Chicago Press, 1977[1959]); Michael Polanyi, *Knowing And Being*. With an Introduction by Marjorie Grene (Chicago: Chicago University Press, 1969); Robert K. Merton, *The Sociology of Science* (Chicago: University of Chicago Press, 1973); Ian I. Mitroff, "Norms and Counter-norms in a Selected Group of Apollo Moon Scientists: A Case Study in the Ambivalence of Scientists." *American Sociological Review* 39 (1974): 579–95; Richard Whitley, *The Intellectual and Social Organization*.

54. Kuhn, *The Structure of Scientific Revolutions*, 226.

55. Donald Pelz and Frank M. Andrews, *Scientists in Organizations (revised edition)* (New York: Wiley, 1976), xv.

56. Keith Sawyer, *Group Genius: The Creative Power of Collaboration* (New York: Basic Books, 2007) 56.

57. Pelz and Frank M. Andrews, *Scientists in Organizations*; Edward J. Hackett, "Essential Tensions: Identity, Control, and Risk in Research," *Social Studies of Science* 35 (2005): 789–826; John N. Parker and Beatrice I. Crona, "On Being All Things to All People: Boundary Organizations and the Contemporary Research University," *Social Studies of Science* 42 (2012): 262–289.

58. Pelz and Frank M. Andrews, *Scientists in Organizations*; Mitroff, "Norms and Counter-norms"; Hackett, "Essential Tensions"; Parker and Hackett, "Hot Spots".

59. Reichenbach, *Experience and Prediction*.

60. Karl Popper, *Conjectures and Refutations* (London: Routledge, 1963).

61. Kuhn, *Essential Tensions*.

62. Gerald Holton, "Subelectrons, Presuppositions and the Millikan-Ehrenhaft Dispute," *Historical Studies in the Physical Sciences* 9 (1978): 166–224.

63. Ernest B. Hook, *Prematurity in Scientific Discovery: On Resistance and Neglect* (Los Angeles: University of California Press, 2002).

64. Parker and Hackett, "Hot Spots".

65. Rhoten, O'Connor and Hackett, "The Act of Collaborative Creation," 290–291.

66. Janis, *Victims of Group Think*.

67. Mullins, *Theories and Theory Groups;* see also Hull 1988 for similar behavior among the contentious cladists and systematists: David L. Hull, *Science as a Process: An Evolutionary Account of the Social and Conceptual Development of Science* (Chicago; University of Chicago Press, 1988).

68. Christopher Henke, *Cultivating Science: Harvesting Power: Science and Industrial Agriculture in California* (Cambridge, MA: MIT Press, 2008).

69. Donald Cash et al., "Knowledge Systems for Sustainable Development," *PNAS* 100 (2003): 8086–8091.

70. Gerald Gordon and Sue Marquis, "Freedom, Visibility of Consequence, and Scientific Innovation." *American Journal of Sociology* 72 (1966): 195–202.

71. Donald Stokes, *Pasteur's Quadrant* (Washington, DC: Brookings, 1997).

72. Warren Hagstrom, *The Scientific Community* (New York: Basic Books, 1965): 106.

73. Hackett, "Essential Tensions," 803.

74. François Jacob, *The Statue Within* (New York: Basic Books, 1995 [1987]): 258).

75. Hackett, "Essential Tensions," 801.

76. Hackett, "Essential Tensions"; Edward J. Hackett and John N. Parker, "Research Groups," in *Leadership in Science and Technology: A Reference Handbook*, ed. William Sims Bainbridge (Thousand Oaks, CA: Sage, 2011), 164–174.

77. Mihaly Csikszentmihalyi, *Creativity: Flow and the Psychology of Discovery and Invention* (New York: Harper Perennial, 1996).

78. Sawyer, *Group Genius;* Corte, *Subcultures and Small Groups.*

79. Farrell, *Collaborative Circles,* 23. Conditions fostering group flow include well-defined but open-ended goals, full and spontaneous engagement, complete concentration, group autonomy, balanced participation, personal familiarity, constant and spontaneous communication, and building on others' ideas: Sawyer, *Group Genius;* Corte, *Subcultures and Small Groups.*

80. Robert Frodeman, Julie Thomson Klein and Carl Mitcham, *The Oxford Handbook of Interdisciplinarity* (Oxford: Oxford University Press, 2010).

81. Collins, Robert Evan and Mike Gorman, "Trading Zones".

82. Barry J. Marshall, "One Hundred Years of Discovery and Rediscovery of Helicobacter pylori and Its Association with Peptic Ulcer Disease," in *Helicobacter pylori: Physiology and Genetics*, ed. Harry LT Mobley, George L Mendz and Stuart L Hazell (Washington, DC: ASM Press, 2001).

83. Mitroff, "Norms and Counter-norms".

84. John A. Stewart, *Drifting Continents and Colliding Paradigms: Perspectives on the Geoscience Revolution* (Bloomington, IN: Indiana University Press, 1990).

85. Holton, "Subelectrons, Presuppositions".

86. Heather E. Douglas, *Science, Policy, and the Value-Free Ideal* (Pittsburgh: University of Pittsburgh Press, 2009).

87. Paul Thagard, *Hot Thought: Mechanisms and Applications of Emotional Cognition* (Cambridge, MA: MIT Press, 2006); Parker and Hackett "Hot Spots"; Parker and Hackett, "The Sociology of Science".

88. Bent Flyvbjerg, *Making Social Science Matter* (Cambridge, England: Cambridge University Press, 2001); Bent Flyvbjerg, "Phronetic Planning Research: Theoretical and Methodological Reflections." *Planning Theory and Practice* 5 (2004): 283–306; Michael J. Sandel, *Justice: What's the Right Thing to Do?* (New York: Farrar, Straus, and Giroux, 2009); Amartya Sen, *The Idea of Justice* (Cambridge, MA: Belknap Press, 2009).

Acknowledgements This research would not have been possible without the cheerful and enduring support of Jim Reichman, Stephanie Hampton, Frank Davis, the NCEAS staff, and hundreds of scientists who took time from their research visits to answer our questions, complete our surveys, explain things to us, and simply allow us to spend time with them. We thank Nancy Grimm for suggesting NCEAS as a research site and Jonathon Bashford for helpful analyses and discussions. An earlier version of some of the ideas and evidence presented in this paper appeared in Hackett et al. (2008).

This work was supported by the National Science Foundation (SBE 98–96330 to Hackett, SBE 1242749 to Hackett and Parker) and by the National Center for Ecological Analysis and Synthesis, Santa Barbara, CA (DEB 94–21535).

We are deeply grateful to Dave Conz for years of collegiality and conviviality, and dedicate this work to him.

METHODOLOGICAL APPENDIX

National Center for Ecological Analysis and Synthesis

Our study of NCEAS began in 1998 and continues to the present. We interviewed administrators, resident scientists, and working group members; examined documents, publications, and citation data; observed working groups; and administered a brief questionnaire. One of us was in residence as a participant observer in 2004–2005, the other from 2008 to 2011.

During those and other visits we spent more than 140 hours in ethnographic observation of working groups, and hundreds more observing informal interaction in the groups and conducting interviews. We observed the entire course of each working group session, arriving at NCEAS each morning before scientists arrived to work, and leaving only after all work had been completed that day. We took detailed notes of group behavior as it occurred, adding detail from recollection during the evening. Throughout the project we have been deeply engaged with the Center: material from our study was used in official evaluative site visits (1999, 2002, and 2008), discussed on several occasions with the NCEAS director, and summarized at length within the Center's (successful) renewal proposal.

The RA

We conducted semi-structured, in-depth interviews with RA members from 2003 to 2010 in Tempe, AZ; Madison, WI; Decatur, GA; Cedar Key, FL; Stockholm, Sweden; Wageningen, Netherlands; Kruger National Park, South Africa; and Gabriola Island, Canada. 30 initial interviews were conducted, each lasting 45–90 minutes. We asked about RA's past, present and future directions, and about practices occurring on island, group structure, organizational and intellectual challenges, intergenerational dynamics, and receptivity of their work by the scientific community. We also inquired into group leadership and selection processes, the successes and failures of specific projects, researchers' personal motivations, interdisciplinary interactions, and the impact of resilience research on science and policy. After the first round of interviewing, initial findings were tested through dozens of follow-up interviews ranging from brief exchanges to multihour conversations. Altogether, we spoke with more than 50 researchers (most several times), including all but two members identified by RA founders as central to its development, and many operating at RA's periphery (junior scientists and new members). We also spoke with many non-RA members about the group.

We conducted ethnographic observations, beginning during the second author's stay as a visiting researcher at Stockholm University's RA node (May–July 2003). Not an RA member, he returns regularly to interview, observe, and trace changes over time. Over 200 hours of (non-participant) observations were undertaken "on island" at Kruger National Park, South Africa (2006; five days and nights), at the first resilience conference (Stockholm University, 2008; four days and two nights), and on Gabriola Island, British Columbia (2009; ten days and nights).

On island, access was provided to all activities except board meetings. Observations began during breakfast and continued late into the evening. Meals were eaten together and drinks shared. Scientific conversations were observed, as were discussions regarding RA's current organization and future directions, and informal activities (safari excursions, limerick contests). Observations at the Stockholm conference included scientific presentations, organizational meetings, science-policy dialogues, and informal activities (e.g. the resilience art exhibit at The Swedish Museum of Natural History).

References

Accademia Nazionale dei Lincei. History. http://www.lincei.it/modules.php?name=Content&pa=showpage&pid=21.html. Accessed 13 Feb 2012

Amabile, Teresa M. 1983. Social psychology of creativity: A componential conceptualization. *Journal of Personality and Social Psychology* 45: 997–1013.

Amabile, Teresa M. 2012. Componential theory of creativity. Harvard Business School working paper 12–096, Harvard Business School.. http://www.hbs.edu/faculty/Publication%20Files/12-096.pdf

Andreae, Johann Valentin. 1916[1619]. *Christianopolis: An idea state of the seventeenth century.* New York: Oxford University Press.

Bacon, Francis. 1627. *The New Atlantis.* London: John Haviland for William Lee.

Bernal, Joseph. 1967 [1939]. *Social functions of science.* Cambridge, MA: M.I.T. Press.

Bierman, Judah. 1963. Science and society in the New Atlantis and other Renaissance Utopias. *Proceedings of the Modern Language Association* 78: 492–500.

Bowden, Edward M., and Mark Jung-Beeman. 2003. Normative data for 144 compound remote associate problems. *Behaviorial Research Methods, Instruments, & Computers* 35: 634–639.

Campanella, Tomasso. 1602. *The city of the Sun.* New York: P.F. Collier & Sons.

Carpenter, Stephen, E. Virginia Armbrust, Peter W. Arzberger, F. Stuart Chapín, James J. Elser, Edward J. Hackett, Anthony R. Ives, Peter M. Kareiva, Mathew A. Leibold, Per Lundberg, Marc Mangel, Nirav Merchant, William W. Murdoch, Margaret A. Palmer, Debra P.C. Peters, Steward T.A. Pickett, Kathleen K. Smith, Diana H. Wall, and Ann S. Zimmerman. 2009. Accelerate synthesis in ecology and environmental sciences. *BioScience* 59: 699–701.

Cash, David W., William C. Clark, Frank Alcock, Nancy M. Dickson, Noelle Eckley, David H. Guston, Jill Jäger, and Ronald B. Mitchell. 2003. Knowledge systems for sustainable development. *PNAS* 100: 8086–8091.

Collins, Randall. 1998. *The sociology of philosophies.* Cambridge, MA: Harvard University Press.

Collins, H.M., and Robert J. Evans. 2007. *Rethinking expertise*. Chicago: University of Chicago Press.

Collins, Harry, Robert Evans, and Michael Gorman. 2007. Trading zones and interactional expertise. *Studies in History and Philosophy of Science* 38: 657–666.

Corte, Ugo. 2012. Subcultures and small groups: A social movement theory approach. PhD dissertation, Uppsala University.

Csikszentmihalyi, Mihaly. 1996. *Creativity: Flow and the psychology of discovery and invention*. New York: Harper Perennial.

DeFries, Ruth, Erle C. Ellis, F. Stuart Chapin III, Pamela A. Matson, B.L. Turner II, Arun Agrawal, Paul J. Crutzen, Chris Field, Peter Gleick, Peter M. Kareiva, Eric Lambin, Diana Liverman, Elinor Ostrom, Pedro A. Sanchez, and James Syvitski. 2012. Planetary opportunities: A social contract for global change science to contribute to a sustainable future. *BioScience* 62: 603–606.

Douglas, Heather E. 2009. *Science, policy, and the value-free ideal*. Pittsburgh: University of Pittsburgh Press.

Ecological Society of America and the Association of Ecosystem Research Centers. 1992/1993. *National Center for Ecological Synthesis: Scientific objectives, structure, and implementation*, Report from a joint committee, based on a workshop held in Albuquerue, NM, 25–27 Oct 1992, 1993.

Farrell, Michael P. 2001. *Collaborative circles: Friendship dynamics and creative work*. Chicago: University of Chicago Press.

Fleck, Ludwik. 1979 [1935]. *Genesis and development of a scientific fact*. Chicago: University of Chicago Press.

Flyvbjerg, Bent. 2001. *Making social science matter*. Cambridge: Cambridge University Press.

Flyvbjerg, Bent. 2004. Phronetic planning research: Theoretical and methodological reflections. *Planning Theory and Practice* 5: 283–306.

Frickel, Scott, and Neil Gross. 2005. A general theory of scientific/intellectual movements. *American Sociological Review* 70: 204–232.

Frodeman, Robert, Julie Thomson Klein, and Carl Mitcham. 2010. *The Oxford handbook of interdisciplinarity*. Oxford: Oxford University Press.

Gordon, Gerald Sue Marquis. 1966. Freedom, visibility of consequence, and scientific innovation. *American Journal of Sociology* 72: 195–202.

Griffith, Belver C., and Nicholas C. Mullins. 1977. Coherent groups in scientific change: 'Invisible colleges' may be consistent throughout science. *Science* 177: 959–964.

Guimerà, Roger, Brian Uzzi, Jarrett Spiro, and Luís A. Nunes Amaral. 2005. Team assembly mechanisms determine collaboration network structure and team performance. *Science* 308: 697–702.

Gunderson, Lance, and C.S. Holling. 2002. *Panarchy: Understanding transformations in human and natural systems*. Washington, DC: Island Press.

Hackett, Edward J. 1990. Science as a vocation in the 1990s: The changing organizational culture of academic science. *Journal of Higher Education* 61: 241–279.

Hackett, Edward J. 2005. Essential tensions: Identity, control, and risk in research. *Social Studies of Science* 35: 789–826.

Hackett, Edward J., and John N. Parker. 2011. Research groups. In *Leadership in science and technology: A reference handbook*, ed. William Sims Bainbridge, 164–174. Thousand Oaks: Sage.

Hackett, Edward J., and John N. Parker. 2016. Ecological science reconfigured: Group and organizational dynamics in scientific change. In *The local configuration of new research fields: On regional and national diversity*, ed. Martina Merz and Philippe Sormani, 153–171. New York: Springer.

Hackett, Edward J., John N. Parker, David Conz, Diana Rhoten, and Andrew Parker. 2008. Ecology transformed: The National Center for Ecological Analysis and Synthesis and the changing patterns of ecological research. In *Scientific collaboration on the Internet*, ed. Gary Olson, Nathan Bos, and Ann Zimmerman, 277–296. Cambridge, MA: MIT Press.

Hagstrom, Warren. 1965. *The scientific community*. New York: Basic Books.

Hampton, Stephanie H., and John N. Parker. 2011. Collaboration and productivity in scientific synthesis. *Bioscience* 61: 900–910.

Heinze, Thomas, Philip Shapira, Juan D. Rogers, and Jacqueline M. Senker. 2009. Organizational and institutional influences on creativity in scientific research. *Research Policy* 38: 610–623.

Henke, Christopher. 2008. *Cultivating science: Harvesting power: Science and industrial agriculture in California*. Cambridge, MA: MIT Press.

Holling, C.S. 1973. Resilience and the stability of ecological systems. *Annual Review of Ecology and Systematics* 4: 1–23.

Hollingsworth, J. Rogers, and Ellen Jane Hollingsworth. 2012. *Fostering scientific excellence: Organizations, institutions, and major discoveries in biomedical science*. New York: Oxford University Press.

Holton, Gerald. 1978. Subelectrons, presuppositions and the Millikan-Ehrenhaft dispute. *Historical Studies in the Physical Sciences* 9: 166–224.

Hook, Ernest B. 2002. *Prematurity in scientific discovery: On resistance and neglect*. Los Angeles: University of California Press.

Hull, David L. 1988. *Science as a process: An evolutionary account of the social and conceptual development of science*. Chicago: University of Chicago Press.

Jacob, François. 1995 [1987]. *The statue within*. New York: Basic Books.

Janis, Irving L. 1972. *Victims of group think*. Boston: Houghton Mifflin Company.

Knorr-Cetina, Karin. 1999. *Epistemic cultures: How the sciences make knowledge*. Cambridge, MA: Harvard University Press.

Kuhn, Thomas S. 1962. *The structure of scientific revolutions*. Chicago: University of Chicago Press.

Kuhn, Thomas S. 1977 [1959]. *Essential tensions: Selected studies in scientific tradition and change*. Chicago: University of Chicago Press.

Levine, John M., and Richard L. Moreland. 2004. Collaboration: The social context of theory development. *Personality and Social Psychology Review* 8: 164–172.

Marshall, Barry J. 2001. One hundred years of discovery and rediscovery of Helicobacter pylori and its association with peptic ulcer disease. In *Helicobacter pylori: Physiology and genetics*, ed. Harry L.T. Mobley, George L. Mendz, and Stuart L. Hazell. Washington, DC: ASM Press.

Mednick, Sarnoff A. 1962. Remote associates test. *Journal of Creative Behavior* 2: 213–214.

Merton, Robert K. 1973. *The sociology of science*. Chicago: University of Chicago Press.

Mitroff, Ian I. 1974. Norms and counter-norms in a selected group of Apollo moon scientists: A case study in the ambivalence of scientists. *American Sociological Review* 39: 579–595.

Mullins, Nicholas C. 1973. *Theories and theory groups in contemporary American sociology*. New York: Harper.

Ornstein, Margaret. 1963. *The role of scientific societies in the seventeenth century*. Hamden: Archon Books.

Parker, John. 2010. Integrating the social into the ecological: Organization and research group challenges. In *Collaboration in the new life sciences*, ed. J.N. Parker, N. Vermeulen, and B. Penders, 85–109. Burlington: Ashgate.

Parker, John N., and Beatrice I. Crona. 2012. On being all things to all people: Boundary organizations and the contemporary research university. *Social Studies of Science* 42: 262–289.

Parker, John N., and Edward J. Hackett. 2012. Hot spots and hot moments in scientific collaboration and social movements. *American Sociological Review* 77: 21–44.

Parker, John N., and Edward J. Hackett. 2014. The sociology of science and emotions. In *Handbook of the sociology of emotions: Volume II*, ed. Jan E. Stets and Jonathan H. Turner, 549–572. New York: Springer.

Pelz, Donald, and Frank M. Andrews. 1976. *Scientists in organizations, revised edition*. New York: Wiley.

Polanyi, Michael. 1969. *Knowing and being. With an introduction by Marjorie Grene*. Chicago: Chicago University Press.

Popper, Karl. 1963. *Conjectures and Refutations*. London: Routledge.

Reichenbach, Hans. 1938. *Experience and prediction: An analysis of the foundations and structure of knowledge*. Chicago: University of Chicago Press.

Rhoten, Diana R., Erin O'Connor, and Edward J. Hackett. 2008. The act of collaborative creation and the art of integrative creativity: Originality, disciplinarity, and interdisciplinarity. *Thesis Eleven* 96: 83–108.

Rockström, Johan, Will Steffen, Kevin Noonel, Åsa Persson, F. Stuart Chapin, Eric F. Lambin, Timothy M. Lenton, Marten Scheffer, Carl Folke, Hans Joachim Schellnhuber, Björn Nykvist, Cynthia A. de Wit, Terry Hughes, Sander van der Leeuw, Henning Rodhe, Sverker Sörlin, Peter K. Snyder, Robert Costanza, Uno Svedin, Malin Falkenmark, Louise Karlberg, Robert W. Corell, Victoria J. Fabry, James Hansen, Brian Walker, Diana Liverman, Katherine Richardson, Paul Crutzen, and Jonathan A. Foley. 2009. A safe operating space for humanity. *Nature* 461: 472–474.

Sandel, Michael J. 2009. *Justice: What's the right thing to do?* New York: Farrar, Straus, and Giroux.

Sawyer, Keith. 2007. *Group genius: The creative power of collaboration.* New York: Basic Books.

Schurman, Rachel, and William Munro. 2006. Ideas, thinkers, and social networks: The process of grievance construction in the anti-genetic engineering movement. *Theory and Society* 35: 1–38.

Sen, Amartya. 2009. *The idea of justice.* Cambridge, MA: Belknap Press.

Shapin, Steven. 2008. *The scientific life: A moral history of a late modern vocation.* Chicago: University of Chicago Press.

Sidlauskas, Brian, Ganeshkumar Ganapathy, Einat Hazkani-Covo, Kristin P. Jenkins, Hilmar Lapp, Lauren W. McCall, Samantha Price, Ryan Scherle, Paula A. Spaeth, and David M. Kidd. 2009. Linking big: The continuing promise of evolutionary synthesis. *Evolution* 64: 871–880.

STEPS Centre. 2010. *Innovation, sustainability, development: A new manifesto.* Sussex: STEPS.

Stewart, John A. 1990. *Drifting continents and colliding paradigms: Perspectives on the geoscience revolution.* Bloomington: Indiana University Press.

Stokes, Donald. 1997. *Pasteur's quadrant.* Washington, DC: Brookings.

Thagard, Paul. 2006. *Hot thought: Mechanisms and applications of emotional cognition.* Cambridge, MA: MIT Press.

Weber, Max. 1946 [1918]. Science as a vocation. In *From Max Weber: Essays in sociology,* ed. Hans H. Gerth and C. Wright Mills, 129–158. New York: Oxford.

Weingart, Peter. 2010. A short history of knowledge formations. In *The handbook of interdisciplinarity,* ed. Robert P. Frodeman, Julie Thompson Klein, and Carl Micham, 3–14. New York: Oxford.

Whitley, Richard. 1984. *The intellectual and social organization of the sciences.* Oxford: Oxford University Press.

Woolley, Anita W., Christopher F. Chabris, Alex Pentland, Nada Hashmi, and Thomas W. Malone. 2010. Evidence for a collective intelligence factor in the performance of human groups. *Science* 330: 686–688.

CHAPTER 4

The Seventh Solvay Conference: Nuclear Physics, Intellectual Migration, and Institutional Influence

Roger H. Stuewer

4.1   ERNEST SOLVAY

No institutional innovation had greater influence on the development of physics prior to the Second World War than the Solvay Conferences in Physics, a significant historical example of the great fruitfulness of investing resources in the exploration of emerging new frontiers in physics (see the introduction of this volume).[1] Their founder, the Belgian industrialist Ernest Solvay, was born in 1838 in Rebecq-Rognon near Brussels where he acquired a modest education in local schools but could not go on to university because of ill health.[2] He therefore entered his father's salt-making business and at age 21 joined his uncle to manage a gasworks in Brussels. Two years later, in 1861, he developed his eponymous process for manufacturing sodium carbonate, produced it in a small plant he built

R.H. Stuewer (✉)
School of Physics and Astronomy, University of Minnesota,
Minneapolis, MN 55455, USA

© The Editor(s) (if applicable) and The Author(s) 2016
T. Heinze, R. Münch (eds.), *Innovation in Science and
Organizational Renewal*, DOI 10.1057/978-1-137-59420-4_4

with his brother Alfred, and then constructed, with financial support from their family, a factory at Couillet, near Charleroi. In 1872, Solvay began to patent every stage of the process and granted licenses to foreign manufacturers, which by 1890 included ones in most Western European countries, Russia, and the USA. By the end of the nineteenth century, Ernest Solvay was a very wealthy man.

Solvay was an autodidact; he read widely, thought deeply about what he read, and put his theories on paper. He proposed a system to explain the entire universe, from the constitution of matter to the organization of human societies. He explained:

> I saw three directions to follow, three problems, that to my mind made up one single problem. The first was a general physics problem: the constitution of matter in time and space—the second was a physiological problem: the mechanics of life, from its most humble manifestations to the phenomenon of thought—and the third was a problem that complemented the first two: the evolution of the individual and of social groups.[3]

Solvay's physics displayed the idiosyncrasy of the autodidact; for example, he argued in his *Gravitique* of 1878 that

> force exists only hypothetically. Movement is neither primordial nor essential to the natural order; on the contrary, it is so completely ruled by gravity, that it seems it only occurs because of gravitational changes.[4]

Solvay enclosed an updated version of his *Gravitique* along with his invitations to Brussels in 1911.

4.2   THE SOLVAY CONFERENCES IN PHYSICS

In 1893, Ernest Solvay created the Solvay Institute for Physiology; in 1902, the Solvay Institute for Sociology; and in 1904, the Solvay School of Commerce, all located in the Scientific Center in Léopold Park in Brussels, and all administered and staffed by a network of close collaborators. In 1910, Solvay contacted one of them, Robert Goldschmidt, Professor of Physical Chemistry at the Free University of Brussels, with the idea of setting up a similar institute in physics. That July, Goldschmidt then asked Walther Nernst, Professor of Physical Chemistry at the University of Berlin, to submit a proposal to Solvay for an international meeting of physicists and chemists to clarify some of the pressing problems in the theories of radiation and specific heats of solids. Solvay agreed to support such a meeting in Brussels, so after Nernst's Berlin colleague Max Planck declined to chair it, he turned to Hendrick Antoon Lorentz, Professor of Theoretical Physics at the University of Leiden—a brilliant choice, since Lorentz was fluent in several languages and was one of the most highly respected theoretical physicists of the period.

The first Solvay Conference in Physics was held at the Solvay Institute for Physiology and in the palatial Hotel Métropole from October 30 to November 3, 1911, on "The Theory of Radiation and Quanta."[5] Eighteen leading physicists and chemists from six European countries were invited, 12 of whom presented talks that were followed by discussions. Also in attendance were three conference secretaries and two representatives of Solvay. Its legendary success inspired Solvay to found a new Institute for Physics, whose broad goal, as defined by Lorentz in early 1912, was to "encourage research intended to extend and above all deepen our knowledge of those natural phenomena in which M. Solvay has a tireless interest."[6] Part of the funds for the new Institute would be set aside for international conferences. It would be managed jointly by a Scientific Committee with Lorentz as its chair and an Administrative Commission responsible for its finances. Solvay imposed a 30-year limit on its funds because, as he prophesied to Lorentz in 1912, "in 30 years from now, physics will have had the last word, civilization will have made progress and we will have a different task to carry out."[7] Neither Solvay nor anyone else could envision that by the end of that 30-year period one horrific world war would have been fought and a second one would have begun.

During the intervening years, however, six more Solvay Conferences in Physics took place, in 1913, 1921, 1924, 1927, 1930, and 1933. Particularly influential were the fifth in 1927 on quantum mechanics[8] and the seventh in 1933 on nuclear physics. Owing to the enduring bitterness following the Great War of 1914–1918, German physicists, including the pacifist Albert Einstein, were excluded from the third and fourth conferences in 1921 and 1924, which prompted the internationalist Lorentz to obtain the approval of Belgian King Albert I in 1926 to nominate Einstein as a member of the Scientific Committee, a symbolic gesture that marked the renewal of ties to German scientists.

4.3 THE SEVENTH SOLVAY CONFERENCE: TOPIC AND PARTICIPANTS

The seventh Solvay Conference provides a clear vantage point from which to view the profound experimental and theoretical developments that were transforming nuclear physics in the fall of 1933: It provided a locus for familiarizing its participants with the new experimental techniques and instruments that were being developed, and it inspired them to create new theoretical ideas and concepts that illuminated the internal structure of the nucleus, all of which they assimilated and then diffused to other physicists in many countries throughout the world, thus injecting entirely new life into the nascent field of nuclear physics at a time when it was being buffeted by strong social and political currents.

The esteemed and beloved Lorentz, who had served as chair of the Scientific Committee for the first five Solvay Conferences, died on February 4, 1928, at the age of 74. Eighteen days later, Paul Langevin, Professor of Physics at the Collège de France, was nominated as Lorentz's successor. Langevin had been invited to every preceding Solvay Conference, and had been a member of the Scientific Committee since the fourth in 1924. He was fluent in English and had a good command of German. Moreover, as the Geneva physicist Charles-Eugène Guye pointed out:

> [He] is also a scientist at the front line, and is fully aware of all the most recent difficulties and problems posed by modern physics. In addition, he has a remarkably clear, precise and quick mind, even when faced with elaborating and analyzing the very thorniest of problems.[9]

Langevin introduced several significant changes into the organization of the Solvay Conferences. First, papers should be submitted a month in advance so that they could be distributed and hence need not be read at the conference. Second, prior to the conference, all papers would be translated into French, English, and German, thus eliminating the need for multilingual secretaries. These requirements meant that the topic and speakers had to be determined 18 months in advance, and that the speakers had to submit their papers one month in advance so that they and the other participants had time to prepare comments on them. Not surprisingly, this schedule often broke down in practice: some speakers submitted their papers only a few days in advance.

The Statutes of the International Solvay Institute of Physics, as adopted in 1930 after earlier modifications, required that the Scientific Committee be composed of eight ordinary members to whom could be added one extraordinary member with the same rights.[10] In addition to Langevin as chair, the current Scientific Committee consisted of Niels Bohr (Copenhagen), Blas Cabrera (Madrid), Peter Debye (Leipzig), Théophile de Donder (Brussels), Albert Einstein (Berlin), Charles-Eugène Guye (Geneva), Abram F. Ioffe (Leningrad), and Owen W. Richardson (London). Langevin convened the Scientific Committee in Brussels in April 1932 to plan the program for the seventh Solvay Conference, to be held 18 months later.

James Chadwick's discovery of the neutron at the Cavendish Laboratory two months earlier was the most exciting new development in physics, so the Scientific Committee did not doubt that nuclear physics was where the "most important problems" lay.[11] They hesitated to choose it as the subject for the seventh Solvay Conference, however, because only six months earlier, in October 1931, Enrico Fermi had organized a major international conference on nuclear physics in Rome,[12] and two years between major conferences on the same subject seemed to be too short a time. Nonetheless, they sensed that an *annus mirabilis* was occurring in nuclear physics: In December 1931, Harold C. Urey at Columbia University had discovered deuterium, and in February 1932, in addition to Chadwick's discovery of the neutron, his colleagues John D. Cockcroft and Ernest T.S. Walton, and Ernest O. Lawrence at the University of California at Berkeley, reported the inventions of their new particle accelerators. Then, in August 1932, Carl D. Anderson at the California Institute of Technology in Pasadena discovered the positron. In the end, the Scientific Committee thus decided to choose nuclear physics as the subject of the seventh Solvay Conference, to be held in Brussels from October 22–29, 1933.

The Scientific Committee decided to invite as many participants as permitted by the Statutes[13]; they differed from those at earlier Solvay Conferences in two important respects. First, Langevin noted, "as we have expressly sought," they were divided equally between experimentalists and theorists "to confront very intimately the efforts of the one with the other." They would bring their diverse areas of expertise to bear in their papers and discussions in an atmosphere of mutual trust, which is crucial in interdisciplinary endeavors.[14] Second, a large number of young physicists were invited. "A young physics," said Langevin, "requires young physicists." "Nothing justifies better our hope in international

Table 4.1   Seventh Solvay Conference participants

Cambridge	**Leningrad**
*Ernest Rutherford (62)	Abram F. Ioffe (53)
**James Chadwick (41)	**George Gamow (29)
*Charles D. Ellis (38)	**Copenhagen**
**John D. Cockcroft (36)	*Niels Bohr (48)
**Paul A.M. Dirac (31)	**Utrecht**
*Ernest T.S. Walton (30)	Hendrik A. Kramers (38)
London	**Rome**
Owen W. Richardson (54)	*Enrico Fermi (32)
*Patrick M.S. Blackett (35)	**Zurich**
Bristol	*Wolfgang Pauli (33)
*Nevill F. Mott (28)	**Madrid**
Manchester	Blas Cabrera (55)
*Rudolf Peierls (26)	**Ghent**
Berlin	Jules E. Verschaffelt (63)
*Lise Meitner (54)	**Liège**
Erwin Schrödinger (46)	Léon Rosenfeld (29)
Leipzig	**Brussels**
*Peter Debye (49)	*Théophile de Donder (61)
**Werner Heisenberg (31)	Édouard Herzen (ca. 56)
Heidelberg	Auguste Piccard (49)
*Walther Bothe (42)	Emile Henriot (48)
Paris	*Ernest Stahel (37)
*Marie Curie (65)	Jacques Errera (37)
*Paul Langevin (61)	Max Cosyns (ca. 30)
*Maurice de Broglie (58)	**Berkeley**
Edmond Bauer (53)	*Ernest O. Lawrence (32)
Louis de Broglie (41)	**Absent**
*Salomon Rosenblum (37)	Paul Ehrenfest (deceased)
**Irène Curie (36)	Albert Einstein (in U.S.A.)
**Frédéric Joliot (33)	Charles-Eugène Guye (ill)
*Francis Perrin (32)	

Note: **Presented paper *Participated in discussion, age in parentheses

collaboration," he declared, than "the appearance in all countries of these young people in whom we place our hope."[15] The result was a "truly international meeting," with 41 participants between the ages of 26 and 65 from 11 countries (Table 4.1). All of the major centers of nuclear research were represented, with the conspicuous exception of the Institut für Radiumforschung in Vienna, whose scientific reputation had suffered greatly in recent years owing to an extended controversy that two physicists there had had with Ernest Rutherford and James Chadwick in Cambridge,[16] leaving the Vienna Institute, in the words of Otto Robert Frisch, as "a sort of *infant terrible* of nuclear physics."[17]

Deeply missed by all was Paul Ehrenfest, who took his own life in Amsterdam on September 25, 1933, just one month before the conference opened. Langevin recalled with a heavy heart Ehrenfest's participation in the third and fifth Solvay Conferences of 1921 and 1927, where he was "so to speak the soul of these meetings."[18] Death had now "destroyed the great spirit and great heart of Ehrenfest," and Langevin considered it to be his "pious duty" to evoke the memory of Ehrenfest, and "to relate how much he will be missed during the course of this meeting."

4.4   INTELLECTUAL MIGRATION

Langevin's words perhaps moved no one more deeply than Ehrenfest's oldest friend at the Solvay Conference, Abram Ioffe from Leningrad, where the two had been colleagues from 1907 to 1912, until Ehrenfest left to become Lorentz's successor at the University of Leiden. Nine years later, Ioffe became director of the Leningrad (then Petrograd) Physico-Technical Institute, where generations of Soviet physicists were educated, one of whom was Ioffe's countryman at the Solvay Conference, George Gamow, who would become part of the greatest intellectual migration in the twentieth century, if not in history.

Gamow began his university studies in Leningrad (then Petrograd) in 1922[19]; he left six years later to spend the summer of 1928 in Max Born's Institute in Göttingen, where he conceived his new quantum-mechanical theory of alpha decay.[20] He then wrote to Niels Bohr, enclosing a letter of reference from Ioffe, proposing to visit Bohr's Institute for Theoretical Physics in Copenhagen before returning to Leningrad. Bohr, however, was so impressed with Gamow and his work after his arrival that he arranged fellowship support for him for the entire 1928–1929 academic year. Subsequently, continuing to transfer his deep knowledge of nuclear physics from one laboratory to another, Gamow spent the 1929–1930 academic year at the Cavendish Laboratory in Cambridge and returned to Copenhagen for the 1930–1931 academic year. He returned to Leningrad that summer, expecting to leave again in the fall to give a paper at Fermi's conference in Rome. This time, however, the Soviet authorities denied him permission to leave, so he remained in Leningrad where he taught physics, married, and made several attempts to escape Russia with his wife. Their plans finally came to fruition when he received an invitation to attend the seventh Solvay Conference, and his wife was mysteriously allowed to accompany him, which to Gamow was (in his inimitable English) "something like a dubble-miracle."[21]

Bohr had persuaded Langevin, who was well known for his communist sympathies and was Chairman of the Franco-Russian Scientific Cooperation Committee,[22] to request the Soviet authorities to officially designate Gamow as a Soviet delegate to the Solvay Conference. Gamow was always uncertain whether Ioffe, as a member of its Scientific Committee, played any role here, because he felt that Ioffe never really liked him very much.[23] In any case, after much uncertainty and "psychologikal warfare," Gamow and his wife were permitted to leave Russia together. After the Solvay Conference, however, they decided not to return, because Gamow knew that political interference in Russia had greatly increased: "proletarian science" was now supposed to combat "erring capitalistic science."[24] He and his wife spent successive two-month periods in Paris, Cambridge, and Copenhagen, after which they crossed the Atlantic in the early summer of 1934 to enable Gamow to participate in the University of Michigan Summer School in Ann Arbor. While there, he received and accepted an offer of a professorship beginning that fall at George Washington University in Washington, DC, thus completing his migration to the USA.

In complete contrast to Gamow's voluntary emigration was the forced expulsion of scientists and other scholars from Germany as a consequence of Adolf Hitler's brutal racial policies. Event followed event with breathtaking rapidity in 1933: Hitler became Chancellor of Germany on January 30; the Reichstag building in Berlin was torched on February 27; the Enabling Act, which empowered the Nazi regime to govern without a constitution for four years, was passed on March 24; the Nazi Civil Service Law went into effect on April 7; and the infamous book burning in a square opposite the University of Berlin (and in many other university cities as well) took place on the evening of May 10—a scene, one observer said, "which had not been witnessed in the Western world since the late Middle Ages."[25] On April 15, a correspondent for the *New York Evening Post* had reported that

an indeterminate number of Jews have been killed. Hundreds of Jews have been beaten or tortured. Thousands of Jews have fled. Thousands of Jews have been, or will be, deprived of their livelihood. All of Germany's 600,000 Jews are in terror.[26]

The Nazi Civil Service Law of April 7 precipitated an unprecedented intellectual migration[27]; on May 19, *The Manchester Guardian* published a list of nearly 200 mostly Jewish scholars who had been dismissed between April 14 and May 4 from over 30 institutions of higher learning throughout Germany.[28] That number would climb significantly in the following months.

To assist the exiled scholars, refugee organizations were established rapidly in Europe and the USA. In England, the Academic Assistance Council was established in early May 1933, with Ernest Rutherford as president. In Denmark, Niels Bohr's Institute became a haven of longer or shorter duration for many refugees from Nazi Germany. In the USA, the Emergency Committee for Aid to Displaced German (later Foreign) Scholars was established and began its work in early June 1933. These rescue efforts were all the more remarkable because of the severe economic depression. The broadcast journalist Edward R. Murrow, second-in-command of the Emergency Committee, noted that by October 1933—just at the time of the seventh Solvay Conference—more than 2000 out of a total of 27,000 teachers had been dropped from the faculties of some 240 colleges and universities in the USA.[29]

By far the most prominent physicist to be caught in the maelstrom in Germany was Albert Einstein, who had been Ehrenfest's closest friend. Langevin only noted at the Solvay Conference that Einstein had left Europe to fulfill a call to the USA.[30] Everyone present, however, knew that this masked Einstein's true fate, for on March 28, 1933, returning from a trip to the USA, Einstein and his wife Elsa disembarked at Antwerp, Belgium, where he resigned from the Prussian Academy of Sciences and then surrendered his German citizenship at the German embassy in Brussels.[31] On April 21, he also severed his ties with the Bavarian Academy of Sciences. Five months later, he left Belgium for England where on October 3 he delivered his first public address in English as the featured speaker at the Royal Albert Hall in London to a packed audience of over 10,000 people. Announcing himself as "a man, a good European and a Jew," he praised the refugee agencies for their work and spoke vigorously in defense of "intellectual and individual freedom," without which "there would be no Shakespeare, Goethe, Newton, Faraday, Pasteur or Lister."[32] He boarded the *Westernland* at Southampton for New York on October 7, arriving ten days later with his wife Elsa, his secretary Helen Dukas, and his collaborator Walther Meyer, to take up an appointment at the Institute for Advanced Study in Princeton. A few days later, Ernest Rutherford, who was in the chair at Einstein's Albert Hall address, left Cambridge for Brussels to attend the seventh Solvay Conference.

Rutherford thus was personally familiar with the painful circumstances surrounding Einstein's absence in Brussels, but he knew that Einstein's case was far from unique. Everyone present, especially every German present, was fully aware of the devastation that had been wrought in Germany. Werner Heisenberg, for example, had written to Bohr from Leipzig on June 30, 1933,[33] reporting that his Solvay lecture was nearly finished, but then adding that he, Max Planck, and Max von Laue were trying, quite likely unsuccessfully, to retain James Franck and Max Born in Germany, leaving "the future completely uncertain." Heisenberg also mentioned his student Felix Bloch, who had not been invited to the Solvay Conference, but whose fate had become intertwined with the lives of many who were.

Thus, Peter Debye, who had been impressed with Bloch as his student at the *Eidgenössische Technische Hochschule* (ETH) in Zurich,[34] took him along when he moved to Leipzig in 1927, where Bloch became Heisenberg's first Ph.D. student, where he met Rudolf Peierls, another of Heisenberg's students, and where he completed his degree in 1928. Bloch then became Wolfgang Pauli's assistant in Zurich (1928–1929), studied further under Hendrik A. Kramers in Utrecht and with A.D. Fokker in Haarlem (1929–1930), returned to Leipzig as Heisenberg's assistant (1930–1931), spent six months with Bohr in Copenhagen (1931–1932), and completed his *Habilitationsschrift* under Heisenberg in Leipzig in the spring of 1932. As a Swiss citizen, he was exempt from the Nazi Civil Service Law of April 7, 1933, but as a Jew and human being, he found that law intolerable, so he quit his position as *Privatdozent*, went home to Zurich that summer, and despite Heisenberg's urging, refused to return to Leipzig. Instead, he was invited to lecture for a few weeks at the Institut Henri Poincaré in Paris where he lived in Langevin's house, and again visited Bohr's Institute in Copenhagen where he received and accepted an offer of a position at Stanford University. Thus, by the time of the seventh Solvay Conference, Bloch had been associated with no less than seven of its participants, Debye, Heisenberg, Peierls, Pauli, Kramers, Bohr, and Langevin, and it would have been natural for them to have Bloch's odyssey on their minds.

Debye remained in Germany as long as possible, and Heisenberg never left, nor would another Solvay participant, Walther Bothe, from the University of Heidelberg. Lise Meitner, protected by her Austrian citizenship, remained at the Kaiser-Wilhelm Institut für Chemie in Berlin-Dahlem until the summer of 1938 when she was spirited out, eventually going to Stockholm, Sweden. Erwin Schrödinger, Planck's successor in Berlin, although not Jewish, was repulsed by the Nazi racial policies, and when the Oxford professor of physics Frederick A. Lindemann visited him in Berlin in April 1933, he expressed his willingness to accept a position at Oxford.[35] Planck was shaken by Schrödinger's decision to leave Berlin, but Heisenberg was simply angry, writing to his mother that Schrödinger had no reason to leave, "since he was neither Jewish nor otherwise endangered."[36] Actually, Schrödinger had been classified as "politically unreliable," a ground for dismissal under the Nazi Civil Service Law. Lindemann informed Schrödinger that he had been elected a Fellow of Magdalen College on October 3; two weeks later, Schrödinger attended a conference in Paris and then went on to Brussels. On October 24, two days after the Solvay Conference opened, the Berlin *Deutsche Zeitung* carried an article regretting the loss of Schrödinger to German science.[37]

Rudolf Peierls, who at age 26 was the youngest physicist invited to the Solvay Conference, was another gifted theoretical physicist permanently lost to Germany. After studying at the Universities of Berlin (1925–1926) and Munich (1926–1928), he received his Ph.D. degree under Heisenberg in Leipzig in July 1929. He then worked for three years as Pauli's assistant at the ETH in Zurich (1929–1932), during which time he also visited Holland, Denmark, and the Soviet Union, where he met his future wife Genia in Odessa in the summer of 1930. He completed his *Habilitationsschrift* under Pauli and then received a Rockefeller Fellowship to go to Rome and Cambridge (fall 1932 to fall 1933). While with Fermi in Rome, he accepted an appointment at the University of Hamburg to begin at Easter 1933, but by then the Nazis were in power, and Peierls declined the Hamburg offer and went to Cambridge instead. By the time he left for Brussels in October, he had received a two-year grant from a refugee organization to support him at the University of Manchester, where another gifted German refugee, Hans A. Bethe, lived with him and his family for a year in a spare room in their house.[38]

The Nazi racial policies thus impinged directly or indirectly on the lives of many of the physicists at the seventh Solvay Conference, not only those from Germany but others as well, for example, Bohr in Copenhagen and Langevin and Frédéric Joliot in Paris, who sheltered refugees both before and after it. No one at the conference, in fact, was left entirely untouched by the plight of the refugees, for by 1933 physics had become a truly international enterprise, with numerous close scientific and personal ties having been forged by traveling fellowships, visiting lectureships and professorships, and international conferences, The first of these on nuclear physics, a *Physikalische Vortragswoche*, was held at the ETH in Zurich from May 20–24, 1931,[39] where six of the 16 lecturers or participants, Gamow, Bothe, Patrick M.S. Blackett, Pauli, Joliot, and Maurice de Broglie, were also present at the seventh Solvay Conference. That fall, at Fermi's conference in Rome from October 11–18, 1931,[40] which Gamow was unable to attend, in addition to Fermi, 12 of the other 42 lecturers or participants, Nevill F. Mott, Charles D. Ellis, Blackett, Owen W. Richardson, Bothe, Debye, Heisenberg, Meitner, Bohr, Léon Rosenfeld, Marie Curie, and Pauli, attended the seventh Solvay Conference. Two years later, the fifth All-Union Conference in Physics was held in Leningrad from September 24–30, 1933,[41] where 4 of its 11 lecturers, Gamow, Joliot, Francis Perrin, and Paul A.M. Dirac, also attended the seventh Solvay Conference. Probably because Gamow refused to return to the Soviet Union after it, his paper was excluded from the published proceedings of the Leningrad conference "for technical reasons."[42] The seventh Solvay Conference therefore became the fourth international conference on nuclear physics in less than two and a half years, and was attended by no less than 17 of the physicists who had attended one or more of the preceding three. The professional and personal bonds they had formed before coming to Brussels had created a sense of community among them at a time of crisis and diaspora.

4.5   NUCLEAR QUESTIONS

Pioneering new instruments and far-reaching experimental results, and compelling new theoretical insights, took center stage at the seventh Solvay Conference. John Cockcroft opened it by discussing his and Ernest Walton's recent experiments with their eponymous accelerator at the Cavendish Laboratory,[43] and in the discussion, Ernest Lawrence described his new 27-inch cyclotron at Berkeley.[44] Marie Curie insightfully remarked that the Cockcroft–Walton reaction, in which protons bombard lithium-7 to produce two alpha particles ($_3Li^7 + _1H^1 \rightarrow 2_2He^4$), was "the first nuclear reaction in which one can verify with precision ... the relation of Einstein between mass and energy."[45] Until then, physicists had simply taken the validity of Einstein's famous relationship, $E = mc^2$, for granted. It had remained inaccessible to precise experimental test until Kenneth Bainbridge at the Bartol Research Foundation of the Franklin Institute in Philadelphia developed a mass spectrograph with which he determined the mass of lithium-7 to high precision. He found in June 1933 that, "Within the probable error of the measurements the equivalence of mass and energy is satisfied."[46] That was what Marie Curie emphasized at the Solvay Conference.

After Cockcroft, first Chadwick and then Irène Curie and her husband Frédéric Joliot presented papers,[47] both of which impinged on a most fundamental issue, the value of the mass of the neutron,[48] which illustrates the positive role that scientific competition plays in the generation of new knowledge.

In May 1932, Chadwick calculated the mass of the neutron from the reaction of alpha particles on boron-11 ($_5B^{11} + _2He^4 \rightarrow _7N^{14} + _0n^1$), finding that it was 1.0067 amu (atomic mass units). That was less than the sum of the proton and electron masses at 1.0078 amu, from which Chadwick concluded that the neutron consists of a proton-electron compound with a binding energy of 1–2 MeV (million electron volts), which supported the model of the neutron that Rutherford had proposed as long ago as 1920. In June 1933, Lawrence challenged Chadwick's value based on experiments he and colleagues had carried out in Berkeley, in which they had bombarded various elements and compounds with deutons (later called deuterons). Lawrence concluded that "the deuton ... is breaking up, presumably into a proton and a neutron," from which reaction he calculated that the mass of the neutron was only about 1.0006 amu,[49] much lower than Chadwick's value of 1.0067 amu.

In July 1933, Curie and Joliot challenged both Chadwick's and Lawrence's values,[50] arguing that when alpha particles bombard boron, they interact not with the heavy isotope of boron, boron-11 ($_5B^{11}$), as Chadwick had assumed, but with the light isotope of boron, boron-10 ($_5B^{10}$), and produce either a neutron and a positron or a proton. They set up the mass–energy equations for these two reactions, subtracted one from the other, and calculated that the mass of the neutron was 1.011 amu, higher than Chadwick's value, and much higher than Lawrence's. Indeed, since the sum of the neutron and positron masses was 1.0115 amu, which exceeds the proton mass at 1.0073 amu, they concluded that the proton consists of a neutron–positron compound with a rather high binding energy.

Chadwick, Lawrence, and Curie and Joliot thus arrived in Brussels with what each regarded as conclusive experimental evidence for their very different values of the mass of the neutron. Lawrence confidently repeated his neutron-mass calculation based on his deuton break-up hypothesis,[51] because he took it to be confirmed by recent experiments at Caltech in Pasadena.[52] No one in Brussels, however, sided with him. Chadwick, in particular, did not budge. He recalculated the mass of the neutron from his above alpha-particle reaction with the heavy isotope of boron, and from a similar alpha-particle reaction with lithium-7 ($_3\text{Li}^7$), both of which he took to confirm his model, that the neutron consists of "an intimate union of a proton with an electron."[53] Curie and Joliot again challenged Chadwick's calculation, and again calculated a neutron mass much higher than Chadwick's value.[54]

This was a fundamental issue, because if Chadwick's or Lawrence's values were correct, the neutron should be a proton–electron compound, while if Curie and Joliot's values were correct, the neutron should be a new elementary particle, and the proton should be a neutron–positron compound. Walther Bothe summarized the situation by noting that the "important question" of whether the neutron or the proton was "the actual elementary particle ... still cannot be answered with certainty."[55] That question could not be left unanswered.

Werner Heisenberg brought this question to center stage in Brussels by presenting a *tour de force* on the structure of the nucleus.[56] His paper constituted a milestone in the history of the liquid-drop model of the nucleus,[57] which George Gamow had proposed in February 1929,[58] and which he had developed quantitatively one year later,[59] finding that a plot of the internal energy E of the alpha particles against the number N of them in the nucleus (the nuclear mass–defect curve) has a distinct minimum in it. Three years later, shortly after Chadwick's discovery of the neutron, Heisenberg introduced the concept of charge exchange as the origin of the nuclear force between a proton and a neutron,[60] which Ettore Majorana revised in early 1933 by introducing a new nuclear force involving the exchange of both charge and spin,[61] thus binding two protons to two neutrons to form an alpha particle, the basic nuclear constituent that Gamow had assumed. Heisenberg succinctly noted that Majorana's theory thus could be "considered as corresponding to a form of Gamow's [liquid-]drop model made precise by the neutron hypothesis."[62] Assuming therefore that nuclei are composed of neutrons and protons, Heisenberg carried out a long quantum-mechanical calculation and again found that the nuclear mass–energy curve has a distinct minimum in it. He therefore provided a new and deeper theoretical foundation for Gamow's liquid-drop model.

In the discussion following Heisenberg's paper, Wolfgang Pauli stole the show by advancing for the first time for publication his neutrino hypothesis in connection with the theory of beta decay.[63] In 1927, Charles D. Ellis and William A. Wooster at the Cavendish Laboratory had proved that the beta particles emitted from RaE ($_{83}Bi^{210}$) have a continuous distribution of energies,[64] a decisive result that Lise Meitner and Wilhelm Orthmann in Berlin confirmed in 1930.[65] This had led Pauli to propose at the end of 1930 the possibility that an electrically neutral spin-1/2 particle of small mass that obeys the exclusion principle is emitted along with an electron in beta decay, thereby preserving the laws of conservation of energy and momentum.[66] Pauli discussed his hypothesis again at a conference in Pasadena, California, in June 1931, in Ann Arbor a few weeks later, and at Fermi's conference in Rome in October 1931, where Samuel Goudsmit remarked on it.[67] By the time of the seventh Solvay Conference, Pauli had adopted Fermi's name, "neutrino," for his hypothetical new particle. Bohr strongly opposed it, arguing instead that the conservation laws were violated in beta decay.[68] Neither Pauli nor Bohr relinquished his position at the Solvay Conference, thus illustrating the strong role that deeply held convictions can play in physics.

Less than two weeks after the close of the Solvay Conference, Heisenberg learned that he had been awarded the Nobel Prize in Physics for 1932, and Dirac and Schrödinger learned that they had been awarded the Nobel Prize in Physics for 1933. They therefore joined five of the other Solvay participants, Marie Curie, Rutherford, Bohr, O.W. Richardson, and Louis de Broglie, and the absent Einstein as Nobel Laureates. That Schrödinger was now in England and Einstein was now in the USA symbolized the intellectual decapitation of Germany that had begun nine months earlier, and would continue apace in the months and years ahead.[69]

Fundamental issues that were raised at the Solvay Conference stimulated the generation of further knowledge in nuclear physics soon after it. In January 1934, Joliot and Curie followed up on the reactions they had reported at the Solvay Conference and discovered artificial radioactivity,[70] the last major discovery in the Institut du Radium in Paris that its founder, Marie Curie, witnessed before her death in July. Also in January 1934, Enrico Fermi adopted Pauli's neutrino hypothesis and published his celebrated theory of beta decay,[71] which provided a firm theoretical basis for excluding electrons from the nucleus. Fermi and his team in Rome then followed up Joliot and Curie's discovery of artificial radioactivity, and in March 1934 demonstrated that slow neutrons readily produce nuclear reactions,[72] a discovery that would have far-reaching consequences in nuclear physics, and for the world.

Also in March 1934, Lawrence in Berkeley found experimentally that his deuton breakup hypothesis had to be abandoned,[73] thus proving that his very low value for the mass of the neutron was in error. Five months later, in August 1934, Chadwick and Maurice Goldhaber, yet another refugee from Nazi Germany, bombarded deuterons (or diplons, as they were called at the Cavendish Laboratory) with energetic gamma rays, breaking them up into neutrons and protons, from which reaction they calculated that the mass of the neutron was 1.0080 amu, greater than Chadwick's earlier value of 1.0067 amu, and even greater than the mass of the hydrogen atom (1.0078 amu).[74] This proved experimentally and conclusively that the neutron was a new elementary particle, and that electrons are not constituents of neutrons, and hence not of nuclei.

These developments all occurred by the time of the fifth international conference on nuclear physics, which took place in London and Cambridge one year after the seventh Solvay Conference, from October 1–6, 1934.[75] The following year, C.F. von Weizsäcker, while working on his *Habilitationsschrift* under Heisenberg in Leipzig, built upon Heisenberg's Solvay analysis and proposed a semiempirical nuclear-mass formula whose plot again displayed the distinct minimum that Gamow and Heisenberg had found earlier.[76] In 1936, after Hans Bethe refined von Weizsäcker's formula somewhat,[77] it became a basic tool for analyzing nuclear binding energies, and thus represented the culmination of the line of development that Gamow had opened up in 1929 with his liquid-drop model of the nucleus. Also in 1936, Niels Bohr initiated another highly productive line of research when he proposed his theory of the compound nucleus to understand nuclear reactions.[78] All of these fruits germinated from experimental and theoretical seeds sown at the seventh Solvay Conference.

4.6   INSTITUTIONAL INFLUENCE

The seventh Solvay Conference was the last one held before Britain and France declared war on Germany on September 3, 1939. By then, the only competing series of international conferences in physics had been arranged by Niels Bohr annually since 1929 at his Institute for Theoretical Physics in Copenhagen, but these were of shorter duration and their proceedings were not published. In 1935, George Gamow adopted Bohr's conferences as a model, enlisted the help of his friends Edward Teller and Merle Tuve, and organized the first Washington Conference on Theoretical Physics. Six more were held by 1941, the most consequential one being the fifth from January 26–28, 1939, where Bohr and Fermi announced the discovery and interpretation of nuclear fission.[79]

After the war, only two more Washington Conferences were held, in 1946 and 1947, but the Solvay Conferences resumed vigorously: 19 more were held from 1948 to 2014 (Table 4.2), bringing the total to 25 by that 50th anniversary of their founding.[80] The smaller, shorter, and elite Washington Conferences, however, served as a closer model for the famous Shelter Island Conference in 1947, and for the succeeding Pocono and Oldstone Conferences in 1948 and 1949. These were precursors to the much larger and more democratic Rochester Conferences that Robert E. Marshak began to organize in 1950, which became the International Conferences on High Energy Physics in 1958,[81] for a total of 36 by 2012.

By contrast, the annual Seven Pines Symposium, which the prominent Minnesota businessman Leland Gohlike founded in 1997 and I organized for the first 11 years, were modeled on the Solvay Conferences in Physics, and hence extended that model to the history and philosophy of physics. I similarly assembled a small advisory committee to determine their topics and to select their speakers and other participants. The first four were held at Gohlike's Seven Pines Lodge in northern Wisconsin (whence the name, the Seven Pines Symposium), while succeeding ones were held at his Outing Lodge at Pine Point near Stillwater, Minnesota. The 20th Seven Pines Symposium was held from May 11–15, 2016 (Table 4.3). Reports were published on the earlier symposia and plans are to publish the proceedings of recent ones. Also like the Solvay Conferences, the Seven Pines Symposia take place over an extended period of time, five days, beginning with cocktails and dinner on Wednesday and closing after lunch on Sunday, with the program on each of the intervening three days consisting of two half-hour talks in the morning and afternoon, each set followed by a half-hour break and an hour and a half discussion, with three hours for lunch and free time, and cocktails and dinner in the evening. In other words, like the Solvay Conference, the entire program provides large segments of time for discussions among the speakers and other participants, a program that has garnered enthusiastic acclaim.

Table 4.2 Solvay Conferences in Physics 1911–2014

1911	The Theory of Radiation and Quanta
1913	The Structure of Matter
1921	Atoms and Electrons
1924	Electric Conductivity of Metals and Related Problems
1927	Electrons and Photons
1930	Magnetism
1933	Structure and Properties of the Atomic Nucleus
1948	Elementary Particles
1951	The Solid State
1954	Electrons in Metals
1958	The Structure and Evolution of the Universe
1961	Quantum Field Theory
1964	The Structure and Evolution of Galaxies
1967	Fundamental Problems in Elementary Particle Physics
1970	Symmetry Properties of Nuclei
1973	Astrophysics and Gravitation
1978	Order and Fluctuations in Equilibrium and Nonequilibrium Statistical Mechanics
1982	Higher Energy Physics
1987	Surface Science
1991	Quantum Optics
1998	Dynamical Systems and Irreversibility
2001	The Physics of Communication
2005	The Quantum Structure of Space and Time
2008	Quantum Theory of Condensed Matter
2011	The Theory of the Quantum World
2014	Astrophysics and Cosmology

Table 4.3 Seven Pines Symposia 1997–2015

1997	Historical Perspectives and Philosophical Problems in the Unification of Physics
1998	Historical and Philosophical Perspectives on the Interplay of Physics and Mathematics
1999	The Field Concept in Physics
2000	Issues in Modern Cosmology
2001	The Quantum Nature of Gravitation, Space, and Time
2002	Symmetry and Symmetry Breaking in Physics
2003	The Concept of the Vacuum in Physics
2004	Quantum Mechanics, Quantum Information, and Quantum Computation
2005	The Classical-Quantum Borderlands
2006	Probability and Improbability in Science
2007	Emergence: From Physics to Biology
2008	The Unseen Universe: Dark Energy and Dark Matter
2009	Effective Field Theories in Condensed Matter Physics
2010	Decoherence and Entanglement
2011	The Origins of Life
2012	Analogy and Duality in Physics
2013	The Conceptual Development of Quantum Physics
2014	The Conceptual Development of Quantum Physics II
2015	General Relativity: A Hundred Years After Its Birth
2016	The Big Questions: Fundamental Problems in Physics

4.7 Conclusions

The Belgian industrialist Ernest Solvay demonstrated the vital role of personal philanthropy by founding and supporting the Solvay Conferences in Physics, the most influential institutional innovation for generating new knowledge in physics prior to the Second World War, and a significant historical example of the great fruitfulness of investing resources to explore emerging new frontiers in physics. Their success rested on the establishment of a Scientific Committee that was chaired by a leading physicist and was comprised of other leading physicists whose purpose was to identify their topics and to select prominent experimental and theoretical physicists to be invited as speakers and other participants. Each conference addressed a pressing issue at the forefront of physics, took place over a period of five or more days, and was held in a hospitable setting conducive to fruitful exchanges of ideas among the participants. That was the heart and soul of the conferences, which could not be achieved through correspondence or other means of scientific communication.

The seventh Solvay Conference in October 1933 took place at a time of political upheaval, especially in Germany following the passage of the Nazi Civil Service Law on April 7, 1933, which produced the most consequential migration of physicists in history and the concomitant diffusion of new knowledge in the field of nuclear physics to many countries of the world. The profound transformations that also took place in experimental and theoretical nuclear physics were no less consequential, examples being the invention of new particle accelerators and the determination of the mass of the neutron.

The influence of the Solvay Conferences in Physics extended not only to other conferences in physics to some extent but also to conferences in the history and philosophy of physics, in particular to the Seven Pines Symposia that were established in 1997.

Notes

1. Jagdish Mehra, *The Solvay Conferences on Physics: Aspects of the Development of Physics since 1911* (Dordrecht: Reidel, 1975); Pierre Marage and Grégoire Wallenborn, *The Solvay Councils and the Birth of Modern Physics* (Basel, Boston, Berlin: Birkhäuser Verlag, 1999).
2. Mehra, *Solvay Conferences*, 1–3; Marage and Wallenborn, *Solvay Councils*, 1–6; W.A. Campbell, "Solvay, Ernest," in Charles Coulston Gillispie, *Dictionary of Scientific Biography* Vol. XII (New York: Charles Scribner's Sons, 1975), 520–521.

3. Quoted in Marage and Wallenborn, *Solvay Councils*, 9.
4. Quoted in ibid.
5. P. Langevin et M. de Broglie, *La Théorie du Rayonnement et les Quanta. Rapports et Discussions de la Réunion tenue á Bruxelles, du 30 octobre au 3 novembre 1911. Sous les Auspices de M. E. Solvay* (Paris, Gauthier-Villars, 1912).
6. Quoted in Marage and Wallenborn, *Solvay Councils*, 18.
7. Quoted in ibid., 19.
8. *Électrons et Photons. Rapports et Discussions du Cinquième Conseil de Physique tenu a Bruxelles du 24 au 29 octobre 1927 sous les Auspices de l'Institut International de Physique Solvay*. Publiés par la Commission administrative de l'Institut (Paris, Gauthier-Villars et Cⁱᵉ, 1928); Guido Bacciagaluppi and Antony Valentini, *Quantum Theory at the Crossroads: Reconsidering the 1927 Solvay Conference* (Cambridge: Cambridge University Press, 2009).
9. Quoted in Marage and Wallenborn, *Solvay Councils*, 41.
10. Institut International de Physique Solvay, *Structure et Propriétés des Noyaux Atomiques. Rapports et Discussions du Septieme Conseil de Physique tenu a Bruxelles du 22 au 29 Octobre 1933* (Paris: Gauthier-Villars, 1934), 348–353, on 348; Roger H. Stuewer, "The Seventh Solvay Conference: Nuclear Physics at the Crossroads," in *No Truth Except in the Details: Essays in Honor of Martin J. Klein*, ed. A.J. Kox and Daniel M. Siegel (Dordrecht, Boston, London: Kluwer, 1995), 333–362, on 333.
11. Langevin, Solvay *Rapports*, ix.
12. Reale Accademia d'Italia, *Convegno di Fisica Nucleare Ottobre 1931-IX* (Roma: Reale Accademia d'Italia, 1932).
13. Langevin, Solvay *Rapports*, x.
14. Robert Crease, "Physical Sciences," in *The Oxford Handbook of Interdisciplinarity*, ed. Robert Frodeman et al. (Oxford/New York: Oxford University Press, 2010), 79–102, especially 96.
15. Langevin, Solvay *Rapports*, x.
16. Roger H. Stuewer, "Artificial Disintegration and the Cambridge-Vienna Controversy," in *Observation, Experiment, and Hypothesis in Modern Physical Science*, ed. Peter Achinstein and Owen Hannaway (Cambridge, MA and London: The MIT Press, 1985), 239–307.
17. Otto R. Frisch, "The Discovery of Fission: How It All Began," *Physics Today* 20 (November 1967), 43–48, on 43–44.
18. Langevin, Solvay *Rapports*, viii.
19. George Gamow, *My World Line: An Informal Autobiography* (New York: Viking, 1970), 28–54.
20. George Gamow, "Zur Quantentheorie des Atomkernes," *Zeitschrift für Physik* 51 (1928), 204–212; Roger H. Stuewer, "Gamow's Theory of

Alpha-Decay," in *The Kaleidoscope of* Science, ed. Edna Ullmann-Margalit (Dordrecht: Reidel, 1986), 147–186.

21. George Gamow, "The Facts Concerning My Getaway from Russia" *Gamow papers* (Library of Congress, October 1950), 2.

22. Gamow, *My World Line*, 128.

23. Gamow, "Facts Concerning My Getaway", 3.

24. Ibid., 1.

25. William L. Shirer, *The Rise and Fall of the Third Reich: A History of Nazi Germany* (New York: Simon and Schuster, 1960), 241.

26. Quoted in Norman Bentwich, *The Refugees from Germany April 1933 to December 1935* (London: Allen & Unwin, 1936), 28–29.

27. Laura Fermi, *Illustrious Immigrants: The Intellectual Migration from Europe 1930–41* (Chicago: University of Chicago Press, 1968), 174–214; Charles Weiner, "A New Site for the Seminar: The Refugees and American Physics in the Thirties," in *The Intellectual Migration: Europe and America, 1930–1960*, ed. Donald Fleming and Bernard Bailyn (Cambridge, MA: Harvard University Press, 1969), 190–234; Daniel J. Kevles, *The Physicists: The History of a Scientific Community in Modern America* (New York: Knopf, 1978), 279–283; Gerald Holton, "The Migration of Physicists to the United States," in *The Muses Flee Hitler: Cultural Transfer and Adaptation 1930–1945*, ed. Jarrell C. Jackman and Carla M. Borden (Washington: Smithsonian Institution Press, 1983), 169–188; Roger H. Stuewer, "Nuclear Physicists in a New World: The Émigrés of the 1930s in America," *Berichte zur Wissenschaftsgeschichte* 7 (1984), 23–40.

28. Reproduced in Weiner, "New Site," 234.

29. Nathan Reingold, "Refugee Mathematicians in the United States, 1933–1941: Reception and Reaction," in Jackman and Borden, *Muses*, 206.

30. Langevin, Solvay *Rapports*, vii.

31. Albert Einstein, *Ideas and Opinions* (New York: Bonanza, 1954), 205–209; Otto Nathan and Heinz Norden, *Einstein on Peace* (New York: Avenel, 1960), 215–216; Ronald W. Clark, *Einstein: The Life and Times* (New York: World, 1971), 463–464.

32. Nathan and Norden, *Einstein on Peace*, 237–238.

33. Heisenberg to Bohr, June 30, 1933, Bohr Scientific Correspondence, Archive for History of Quantum Physics (AHQP).

34. This account is based on two interviews with Bloch, the first one with Thomas S. Kuhn, May 14, 1964 (AHQP), the second one with Charles Weiner, August 15, 1968, Center for History of Physics, American Institute of Physics, College Park, Maryland.

35. Walter Moore, *Schrödinger: life and thought* (Cambridge: Cambridge University Press, 1989), 269.

36. Quoted in David C. Cassidy, *Uncertainty: The Life and Science of Werner Heisenberg* (New York: Freeman, 1992), 310.

37. Moore, *Schrödinger*, 276.

38. Rudolf Peierls, *Bird of Passage: Recollections of a Physicist* (Princeton: Princeton University Press, 1985), 16–81, 90, 99.

39. Egon Bretscher and Eugen Guth, "Zusammenfassender Bericht über die Physikalische Vortragswoche der Eidg. Technischen Hochschule Zürich vom 20–24. Mai 1931," *Physikalische Zeitschrift* 32 (1931): 649–674.

40. Reale Accademia d'Italia, *Convegno di Fisica Nucleare*.

41. M.P. Bronshtein et al. *Atomic Nuclei: A Collection of Papers for the All-Union Nuclear Conference* (Leningrad and Moscow: State Technical-Theoretical Publishing House, 1934) [in Russian]. I am grateful to Morton Hamermesh for translating substantial portions of these proceedings for me.

42. Ibid., Preface.

43. J.D. Cockcroft, "La Désintégration des Éléments par des Protons accélérés," *Solvay Rapports*, 1–56.

44. Ibid., 61–70.

45. Ibid., 76.

46. Kenneth T. Bainbridge, "The Equivalence of Mass and Energy," *Physical Review* 44 (1933), 123.

47. J. Chadwick, "Diffusion anomale des Particules α. Transmutation des Éléments par des Particules α. Le Neutron," *Solvay Rapports*, 81–112; Frédéric and Irène Joliot, "Rayonnement pénétrant des Atomes sous l'Action des Rayons α," ibid., 121–156.

48. Roger H. Stuewer, "Mass-Energy and the Neutron in the Early Thirties," *Science in Context* 6 (1993), 195–238.

49. M.S. Livingston, Malcolm C. Henderson and E.O. Lawrence, "Neutrons from Deutons and the Mass of the Neutron," *Physical Review* 44 (1933), 781–782; quote on 782.

50. Irène Curie and Frédéric Joliot, "La complexité du proton et la masse du neutron," *Comptes rendus hebdomadaries des séances de l'Académie des Sciences* 197 (1933), 237–238.

51. Lawrence, *Solvay Rapports*, 67–69.

52. H. Richard Crane, Charles C. Lauritsen and A. Soltan, "Production of Neutrons by High Speed Deutons," *Physical Review* 44 (1933), 692–693.

53. Chadwick, Solvay *Rapports*, 102.

54. Curie, Joliot, 156.

55. Walther Bothe, "Das Neutron und das Positron," *Die Naturwissenschaften* 21 (1933), 825–831; on 830.

56. Werner Heisenberg, "Considérations théoriques générales sur la Structure du Noyau," *Solvay Rapports,* 289–323.

57. Roger H. Stuewer, "The Origin of the Liquid-Drop Model and the Interpretation of Nuclear Fission," *Perspectives on Science* 2 (1994), 39–92.

58. George Gamow, "Discussion on the Structure of Atomic Nuclei," *Proceedings of the Royal Society of London* [A] 123 (1929), 386–387.

59. George Gamow, "Mass Defect Curve and Nuclear Constitution," *Proceedings of the Royal Society of London* [A] 126 (1930), 632–644.

60. Werner Heisenberg, "Über den Bau der Atomkerne I," *Zeitschrift für Physik* 77 (1932), 1–11; Werner Heisenberg, "Über den Bau der Atomkerne II," *Zeitschrift für Physik* 78 (1932), 156–164; Werner Heisenberg, "Über den Bau der Atomkerne III," *Zeitschrift für Physik* 80 (1933), 587–596.

61. Ettore Majorana, "Über die Kerntheorie," *Zeitschrift für Physik* 82 (1933), 137–145.

62. Heisenberg, *Solvay Rapports,* 316.

63. Pauli, ibid., 324–325.

64. C.D. Ellis and W.A. Wooster, "The Average Energy of Disintegration of Radium E," *Proceedings of the Royal Society of London* [A] 117 (1927), 109–123.

65. Lise Meitner, Wilhelm Orthmann, "Über eine absolute Bestimmung der Energie der primären β-Strahlen von Radium E," *Zeitschrift für Physik* 60 (1930), 143–155.

66. Pauli to Meitner, Geiger and others, December 4, 1930, in *Wissenschaftlicher Briefwechsel mit Bohr, Einstein, Heisenberg u.a.,* Vol. II, ed. Karl von Meyenn (Berlin: Springer, 1985), 39.

67. S. Goudsmit, "Present Difficulties in the Theory of Hyperfine Structure," in *Convegno di Fisica Nucleare,* ed. Reale Accademia d'Italia, 41.

68. Pauli, *Solvay Rapports,* 324; Bohr, ibid., 327–328.

69. The felicitous term, "intellectual decapitation," is Alan D. Beyerchen's; see Jackman and Borden, *Muses,* 41.

70. Irène Curie and Frédéric Joliot, "Un nouveau type de radioactivité," *Comptes rendus* 198 (1934), 254–256; Roger H. Stuewer, "The Discovery of Artificial Radioactivity," in *Oeuvre et engagement de Frédéric Joliot-Curie,* ed. Monique Bordry and Pierre Radvanyi (Les Ulis: EDP Sciences, 2001), 11–20.

71. Enrico Fermi, "Versuch einer Theorie der β-Strahlen. I," *Zeitschrift für Physik* 88 (1934), 161–177.

72. Enrico Fermi, "Radioattività indotta da Bombardamento di Neutroni, I," *Ricerca Scientifica* 5 (1934), 283.

73. Gilbert N. Lewis et al., "On the Hypothesis of the Instability of the Deuton," *Physical Review* 45 (1934), 497.

74. James Chadwick and Maurice Goldhaber, "A 'Nuclear Photo-effect': Disintegration of the Diplon by γ-Rays," *Nature* 134 (1934), 237–238; Maurice Goldhaber, "The Nuclear Photoelectric Effect and Remarks on Higher Multipole Transitions: A Personal History," in *Nuclear Physics in Retrospect: Proceedings of a Symposium on the 1930s*, ed. Roger H. Stuewer (Minneapolis: University of Minnesota Press, 1979), 83–106, on 85–88.

75. The Physical Society, *International Conference on Physics London 1934: A Joint Conference organized by the International Union of Pure and Applied Physics and The Physical Society. Papers & Discussions*, Vol. I (Cambridge: Cambridge University Press, 1935).

76. Carl Friedrich von Weizsäcker, "Zur Theorie der Kernmassen," *Zeitschrift für Physik* 96 (1935), 431–458.

77. Hans A. Bethe and Robert F. Bacher, "Nuclear Physics: A. Stationary States of Nuclei," *Reviews of Modern Physics* 8 (1936): 165–168.

78. Niels Bohr, "Neutron Capture and Nuclear Constitution," *Nature* 137 (1936): 344–348.

79. Roger H. Stuewer, "Bringing the News of Fission to America," *Physics Today* 38 (October 1985), 48–56.

80. "Solvay Conference," website https://en.wikipedia.org/wiki/Solvay_ Conference (accessed August 17, 2014).

81. Silvan S. Schweber, *QED and the Men Who Made It: Dyson, Feynman, Schwinger, and Tomonaga* (Princeton: Princeton University Press, 1994), 179.

Acknowledgments I gratefully acknowledge the supplemental travel support I received from the University of Minnesota Office of the Vice President for Research and the University of Minnesota Retirees Association to attend the "International Conference on Intellectual and Institutional Innovation in Science" at the Berlin-Brandenburg Academy of Sciences and Humanities, September 13–15, 2012, where I presented an early version of this paper.

REFERENCES

Bacciagaluppi, Guido, and Antony Valentini. 2009. *Quantum theory at the crossroads: Reconsidering the 1927 Solvay Conference*. Cambridge: Cambridge University Press.

Bainbridge, Kenneth T. 1933. The equivalence of mass and energy. *Physical Review* 44: 123.

Bentwich, Norman. 1936. *The refugees from Germany April 1933 to December 1935*. London: Allen & Unwin.

Bethe, Hans A., and Robert F. Bacher. 1936. Nuclear physics: A stationary states of nuclei. *Reviews of Modern Physics* 8: 165–168.

Bohr, Niels. 1936. Neutron capture and nuclear constitution. *Nature* 137: 344–348.

Bothe, Walther. 1933. Das Neutron und das Positron. *Die Naturwissenschaften* 21: 825–831.

Bretscher, Egon, and Eugen Guth. 1931. Zusammenfassender Bericht über die Physikalische Vortragswoche der Eidg. Technischen Hochschule Zürich vom 20.–24. Mai 1931. *Physikalische Zeitschrift* 32: 649–674.

Bronshtein, M.P., V.M. Dukelsky, D.D. Ivanenko, and Yu.B. Khariton. 1934. *Atomic nuclei: A collection of papers for the All-Union Nuclear Conference.* Leningrad/Moscow: State Technical-Theoretical Publishing House.

Campbell, W.A. 1975. Solvay, Ernest. In *Dictionary of scientific biography*, vol. XII, ed. Charles Coulston Gillispie, 520–521. New York: Charles Scribner's Sons.

Cassidy, David C. 1992. *Uncertainty: The life and science of Werner Heisenberg.* New York: Freeman.

Chadwick, James. 1934. Diffusion anomale des Particules α. Transmutation des Éléments par des Particules α. Le Neutron. Institut International de Physique Solvay: 81–112.

Chadwick, James, and Maurice Goldhaber. 1934. A 'nuclear photo-effect': Disintegration of the diplon by γ-rays. *Nature* 134: 237–238.

Clark, Ronald W. 1971. *Einstein: The life and times.* New York: World.

Cockcroft, J.D. La Désintégration des Éléments par des Protons accélérés. Institut International de Physique Solvay: 1–56.

Crane, H. Richard, Charles C. Lauritsen, and Andrzej Soltan. 1933. Production of neutrons by high speed deutons. *Physical Review* 44: 692–693.

Crease, Robert. 2010. Physical sciences. In *The Oxford handbook of interdisciplinarity*, ed. Robert Frodeman, Julie Thompson Klein, and Carl Mitcham, 79–102. Oxford/New York: Oxford University Press.

Curie, Irène, and Frédéric Joliot. 1933. La complexité du proton et la masse du neutron. *Comptes rendus hebdomadaries des séances de l'Académie des Sciences* 197: 237–238.

Curie, Irène, and Frédéric Joliot. 1934. Un nouveau type de radioactivité. *Comptes Rendus* 198: 254–256.

Einstein, Albert. 1954. *Ideas and opinions.* New York: Bonanza.

Électrons et Photons. Rapports et Discussions du Cinquième Conseil de Physique tenu a Bruxelles du 24 au 29 octobre 1927 sous les Auspices de l'Institut International de Physique Solvay. 1928. Publiés par la Commission administrative de l'Institut. Paris, Gauthier-Villars et Cⁱᵉ.

Ellis, C.D., and W.A. Wooster. 1927. The average energy of disintegration of radium E. *Proceedings of the Royal Society of London A* 117: 109–123.

Fermi, Enrico. 1934a. Versuch einer Theorie der β-Strahlen. I. *Zeitschrift für Physik* 88: 161–177.

Fermi, E. 1934b. Radioattività indotta da Bombardamento di Neutroni. I. *Ricerca Scientifica* 5: 283.

Fermi, Laura. 1968. *Illustrious immigrants: The intellectual migration from Europe 1930–41*. Chicago: University of Chicago Press.

Frisch, Otto R. 1967. The discovery of fission: How it all began. *Physics Today* 20: 43–48.

Gamow, George. 1928. Zur Quantentheorie des Atomkernes. *Zeitschrift für Physik* 51: 204–212.

Gamow, George. 1929. Discussion on the structure of atomic nuclei. *Proceedings of the Royal Society of London A* 123: 386–387.

Gamow, George. 1930. Mass defect curve and nuclear constitution. *Proceedings of the Royal Society of London A* 126: 632–644.

Gamow, George. 1950. The facts concerning my getaway from Russia... Gamow papers, Library of Congress.

Gamow, George. 1970. *My world line: An informal autobiography*. New York: Viking.

Goldhaber, Maurice. 1979. The nuclear photoelectric effect and remarks on higher multipole transitions: A personal history. In *Nuclear physics in retrospect: Proceedings of a symposium on the 1930s*, ed. Roger H. Stuewer, 83–106. Minneapolis: University of Minnesota Press.

Goudsmit, Samuel A. 1932. Present difficulties in the theory of hyperfine structure. In *Convegno di Fisica Nucleare*, ed. Reale Accademia d'Italia, 41.

Heisenberg, Werner. 1932a. Über den Bau der Atomkerne I. *Zeitschrift für Physik* 77: 1–11.

Heisenberg, Werner. 1932b. Über den Bau der Atomkerne II. *Zeitschrift für Physik* 78: 156–164.

Heisenberg, Werner. 1933. Über den Bau der Atomkerne III. *Zeitschrift für Physik* 80: 587–596.

Heisenberg, Werner. 1934. Considérations théoriques générales sur la Structure du Noyau. Institut International de Physique Solvay: 289–323.

Heisenberg, Werner. Letter to Bohr, June 30, 1933. Bohr Scientific Correspondence, Archive for History of Quantum Physics (AHQP).

Holton, Gerald. 1983. The migration of physicists to the United States. In *The Muses flee Hitler: Cultural transfer and adaptation 1930–1945*, ed. Jarrell C. Jackman and Carla M. Borden, 169–188. Washington: Smithsonian Institution Press.

Institut International de Physique Solvay. 1934. *Structure et Propriétés des Noyaux Atomiques. Rapports et Discussions du Septieme Conseil de Physique tenu a Bruxelles du 22 au 29 Octobre 1933*. Paris: Gauthier-Villars.

Joliot, Frédéric, and Irène Joliot. 1934. Rayonnement pénétrant des Atomes sous l'Action des Rayons α. Institut International de Physique Solvay: 121–156.

Kevles, Daniel J. 1978. *The physicists: The history of a scientific community in Modern America*. New York: Knopf.

Langevin, P., and M. de Broglie. 1912. *La Théorie du Rayonnement et les Quanta. Rapports et Discussions de la Réunion tenue á Bruxelles, du 30 octobre au 3 novembre 1911. Sous les Auspices de M. E. Solvay.* Paris: Gauthier-Villars.

Lewis, Gilbert N., M. Stanley Livingston, Malcolm C. Henderson, and Ernest O. Lawrence. 1934. On the hypothesis of the instability of the deuton. *Physical Review* 45: 497.

Livingston, M.S., Malcolm C. Henderson, and E.O. Lawrence. 1933. Neutrons from deutons and the mass of the neutron. *Physical Review* 44: 781–782.

Majorana, Ettore. 1933. Über die Kerntheorie. *Zeitschrift für Physik* 82: 137–145.

Marage, Pierre, and Grégoire Wallenborn. 1999. *The Solvay Councils and the birth of modern physics*. Basel/Boston/Berlin: Birkhäuser Verlag.

Mehra, Jagdish. 1975. *The Solvay Conferences on physics: Aspects of the development of physics since 1911*. Dordrecht: Reidel.

Meitner, Lise, and Wilhelm Orthmann. 1930. Über eine absolute Bestimmung der Energie der primären β-Strahlen von Radium E. *Zeitschrift für Physik* 60: 143–155.

Moore, Walter. 1989. *Schrödinger: Life and thought*. Cambridge: Cambridge University Press.

Nathan, Otto, and Heinz Norden. 1960. *Einstein on peace*. New York: Avenel.

Pauli. 1985. Letter to Meitner, Geiger, and others, December 4, 1930. In *Wissenschaftlicher Briefwechsel mit Bohr, Einstein, Heisenberg u.a.* vol. II, ed. Karl von Meyenn. Berlin: Springer.

Peierls, Rudolf. 1985. *Bird of passage: Recollections of a physicist*. Princeton: Princeton University Press.

Reale Accademia d'Italia. 1932. *Convegno di Fisica Nucleare Ottobre 1931-IX*. Roma: Reale Accademia d'Italia.

Reingold, Nathan. 1983 Refugee mathematicians in the United States, 1933–1941: Reception and reaction. In *The Muses Flee Hitler: Cultural transfer and adaptation 1930–1945*, ed. Jarrell C. Jackman and Carla M. Borden, 206, Washington: Smithsonian Institution Press.

Schweber, Silvan S. 1994. *QED and the men who made it: Dyson, Feynman, Schwinger, and Tomonaga*. Princeton: Princeton University Press.

Shirer, William L. 1960. *The rise and fall of the Third Reich: A history of Nazi Germany*. New York: Simon and Schuster.

Stuewer, Roger H. 1984. Nuclear physicists in a New World: The emigrés of the 1930s in America. *Berichte zur Wissenschaftsgeschichte* 7: 23–40.

Stuewer, Roger H. 1985a. Bringing the news of fission to America. *Physics Today* 38: 48–56.

Stuewer, Roger H. 1985b. Artificial disintegration and the Cambridge-Vienna controversy. In *Observation, experiment, and hypothesis in modern physical science*, ed. Peter Achinstein and Owen Hannaway, 239–307. Cambridge, MA/ London: The MIT Press.

Stuewer, Roger H. 1986. Gamow's theory of alpha-decay. In *The kaleidoscope of science*, ed. Edna Ullmann-Margalit, 147–186. Dordrecht: Reidel.

Stuewer, Roger H. 1993. Mass-energy and the neutron in the early thirties. *Science in Context* 6: 195–238.

Stuewer, Roger H. 1994. The origin of the liquid-drop model and the interpretation of nuclear fission. *Perspectives on Science* 2: 39–92.

Stuewer, Roger H. 1995. The Seventh Solvay Conference: Nuclear physics at the crossroads. In *No truth except in the details: Essays in honor of Martin J. Klein*, ed. A.J. Kox and Daniel M. Siegel, 333–362. Dordrecht/Boston/London: Kluwer.

Stuewer, Roger H. 2001. The discovery of artificial radioactivity. In *Oeuvre et engagement de Frédéric Joliot-Curie*, ed. Monique Bordry and Pierre Radvanyi, 11–20. Les Ulis: EDP Sciences.

The Physical Society. 1935. *International Conference on Physics London 1934: A joint conference organized by the International Union of Pure and Applied Physics and the Physical Society*, Papers & discussions, vol. I. Cambridge: Cambridge University Press.

von Weizsäcker, Carl Friedrich. 1935. Zur Theorie der Kernmassen. *Zeitschrift für Physik* 96: 431–458.

Weiner, Charles. 1969. A new site for the seminar: The refugees and American physics in the thirties. In *The intellectual migration: Europe and America, 1930–1960*, ed. Donald Fleming and Bernard Bailyn, 190–234. Cambridge, MA: Harvard University Press.

Wikipedia. Solvay Conference. https://en.wikipedia.org/wiki/Solvay_Conference. Accessed 10 May 2016.

CHAPTER 5

"Preservation of the Laboratory Is Not a Mission." Gradual Organizational Renewal in National Laboratories in Germany and the USA

Olof Hallonsten and Thomas Heinze

5.1   INTRODUCTION

The scientific utilization of very large and costly infrastructure—often referred to as "Big Science"—originated with the rise of competition between superpowers at the end of World War II and the tremendous belief in (and fear of) nuclear energy that fed into it. The demonstration of the force of nuclear energy over Hiroshima on August 6, 1945, was essentially the motivation for the initial creation of Big Science laboratories. Generously sponsored national programs for science and technology fostered the development of weapons technologies and civilian use of nuclear

O. Hallonsten
Lund University, Lund, Sweden

T. Heinze (✉)
University of Wuppertal, Wuppertal, Germany

© The Editor(s) (if applicable) and The Author(s) 2016   117
T. Heinze, R. Münch (eds.), *Innovation in Science and Organizational Renewal*, DOI 10.1057/978-1-137-59420-4_5

energy, foremost in the USA, the Soviet Union, Great Britain, and France. The construction of ever-larger particle accelerators to discover new sub-nuclear particles and forces became a manifest feature of the postwar mobilization of science and technology for the benefit of society, the economy, and national security.[1]

The Big Science facilities that were created during this era were essentially mission oriented, and their rise to preeminence in national R&D systems was guided by the unarticulated principle that the accelerators would no longer be useful once the atom's inner structure was fully mapped. To some extent, this premise was correct, since most of the accelerators that were built to search for elementary particles have been shut down. Nowadays, global experimental particle physics (PP) research is concentrated at CERN (Conseil Européen pour la Recherche Nucléaire, the European Organization for Nuclear Research) in Geneva, which hosts the collaborative work of the countries of Europe as well as China, Japan, Russia, and the USA. But interestingly, even following the accomplishment of missions and subsequent desertion of accelerators, the Big Science organizations hosting them remain in place, with very few exceptions, and their shares of national R&D budgets remain as large as ever.

In this chapter, we analyze this seemingly paradoxical state of affairs and explain the organizational processes of change and adaptation that have led to the renewal and survival of Big Science laboratories beyond the completion of their original research missions. In this way, this chapter contributes to what the editors of this volume call "investments in exploration" via adaptation and internal change of existing research organizations. We focus on two systems of national laboratories: that in Germany and that in the USA.[2] Each system functions within its national R&D system to orchestrate the construction and operation of costly research infrastructure and to conduct large-scale scientific and technological programs. Furthermore, both the German and US systems have continued these operations despite considerable changes in the technical nature and areas of use of their infrastructures, and in the contents of their R&D programs, due to the altered demands and expectations from a wide range of scientific fields and from policy makers and society. Important to note is that although Germany and the USA differ fundamentally in the structures of their respective R&D systems, not to mention their (twentieth century) histories and thus their political and institutional foundations for publicly sponsored R&D, the two systems of national laboratories under study are quite alike. As the chapter will show, not least do the processes of adaptation, renewal, and change in the two systems in the past several decades show remarkable similarities. At first sight, therefore, the differences may give the impression of an imbalanced historical comparison of renewal of Big Science in one postwar military and economic superpower and one war-torn European country, but since the two systems under study have far-reaching similarities, the specific combination of Germany and the USA as empirical foci of the analysis adds strength and generalizability to the conclusions.

We examine case studies of two laboratories: DESY (Deutsches Elektronen-Synchrotron, German Electron Synchrotron) in Hamburg, and SLAC (SLAC National Accelerator Laboratory, formerly Stanford Linear Accelerator Center) in Menlo Park, California. These laboratory histories suggest typical patterns according to which laboratories can renew themselves in order to adapt to change. These two laboratories have, in the course of their approximately 50-year histories, undergone gradual but cumulative change with respect to their research missions, from being the flagship PP labs of their respective countries and thus charged with a single mission, to a situation today where they operate no PP machines but rather state-of-the-art photon science (PS) facilities for users from a wide range of the natural sciences, mostly within materials science and the life sciences (broadly defined) but also several other areas. To some extent, these transformations of DESY and SLAC from PP to PS mirror a global development whereby PP has gradually stood back as the main area of utility of large accelerator complexes, and whereby the use of synchrotron radiation (SR) has partly taken its place in contemporary Big Science. Given their sizes, DESY and SLAC have been major players in this global transformation and in some instances pioneered the use of accelerators for SR,[3] but they have not been the lone drivers of the change. Several other interesting studies of the explosive growth of SR as an experimental technique for a wide range of natural sciences exist that use other cases and tell partly different stories.[4] As a dual case study, the chapter therefore has auxiliary relevance as a component piece in the study of how SR came to be a prominent feature of contemporary experimental natural science. But importantly, the focus of the chapter does not lie there, but on the topic of scientific and organizational renewal of national laboratories, and the aims of the chapter are to analyze such renewal in a more general to draw broader conclusions. As mentioned, the complementarity offered by the differences between the two national laboratory systems wherein the cases under study are located adds to the generalizability of the analysis and the conclusions—theoretically, methodologically, and empirically.

Building on prior work, we distinguish four different processes of renewal, and discuss how they interact at the micro (laboratory components) and meso (laboratory) levels and what this means at the macro (research system) level. From this analysis, we infer that the multidimensional and multilevel renewal of national laboratory systems has been instrumental to their survival. The multidimensional renewal processes is key to understanding what the editors of this volume call "investments in exploration." The two cases of DESY and SLAC show how Big Science laboratories were restructured in order to address new scientific problems and challenges. That these investments in SR research/PS have already born fruit, is illustrated by several Nobel Prizes in Chemistry since the late 1990s that have built directly on experimental work at labs like DESY and SLAC, including John Walker (1997), Roderick MacKinnon (2003), Roger Kornberg (2006), Ada Yonath, Thomas Steitz and Venkatraman Ramakrishnan (2009), and Robert Lefkowitz and Brian Kobilka (2012).

We begin by briefly outlining the histories of the two national laboratory systems and how they have grown and transformed since their inception in the late 1940s (USA) and the mid-1950s (Germany). Thereafter, we present the conceptual framework and use it to analyze changes at different levels, considering our knowledge about both the micro and meso levels. The two selected cases enable us to suggest patterns of renewal at the level of the construction and operation of large scientific infrastructure, as well as the scientific activities inside the laboratories. We conclude by focusing on the macro level and the general question of renewal and how laboratories and the systems they comprise have survived despite fundamentally altered political, economic, and military framework conditions.[5]

5.2   SYSTEMS OF NATIONAL LABORATORIES IN THE USA AND GERMANY

The basic purpose of the present analysis is to determine why none of the national laboratories in Germany and the USA have ever been closed, despite considerable changes or even decline and expiration of their original missions. Tables 5.1 and 5.2 list the laboratories of the two systems, along with some basic information.

Ten US National Laboratories are defined as laboratories under the stewardship and main sponsorship of the US Department of Energy's

(DOE) Office of Science.[6] Each is governmentally owned and contractor operated (a legal status commonly called GOCO). Constituting a de facto fourth regular sector of R&D performers in the USA—besides industry, academia, and the government itself—the US National Laboratories are nonprofit but may assume whatever organizational form the contractor finds suitable, including firm, university department, trust, fund, association, or subsidiary and branch of any of these.[7] In addition to these ten laboratories under the Office of Science, there are seven other National Laboratories with responsibility for weapons programs and other classified governmental R&D (including, but not limited to, nuclear arms), which are overseen by other branches of the DOE and, in some cases, the Department of Defense. These seven laboratories are excluded from the analysis since their activities and organizations are classified.

The German Helmholtz Research Centers are wholly civilian R&D centers that operate as limited companies or public and private foundations, and that are all under the umbrella organization the Helmholtz Association. Similar to the situation in the USA, the German Federal Government assigns the Hemholtz centers a unique role in the national R&D system—namely, the construction, maintenance, and operation of large scientific infrastructure.[8] In contrast to the situation in the USA, the Helmholtz Association is a separate legal entity and constitutes an umbrella organization.[9]

Table 5.1 The United States National Laboratories under the DOE Office of Science

Name	Location	Founded
Lawrence Berkeley National Laboratory (LBNL)	Berkeley, CA	1931/1947[a]
Oak Ridge National Laboratory (ORNL)	Oak Ridge, TN	1943/1947[a]
Argonne National Laboratory (ANL)	Argonne, IL	1947
Ames Laboratory (AL)	Ames, IA	1947
Brookhaven National Laboratory (BNL)	Upton, Long Island, NY	1947
Princeton Plasma Physics Laboratory (PPPL)	Princeton, NJ	1953
SLAC National Accelerator Laboratory (SLAC)	Menlo Park, CA	1962
Pacific Northwest National Laboratory (PNNL)	Richland, WA	1965
Fermi National Accelerator Laboratory (FNAL)	Batavia, IL	1967
Thomas Jefferson National Accelerator Facility (TJNAL)	Newport News, VA	1984

[a]These labs were founded in other shapes before (LNBL) and during (ORNL) World War II, and were made National Laboratories in 1947.

Table 5.2   The German Helmholtz Research Centers

Name	Location	Founded
Center for Materials and Coastal Research (GKSS)	Geesthacht/Teltow	1956
Forschungszentrum Jülich (FZJ)	Jülich	1956
Karlsruhe Institute of Technology (KIT)	Karlsruhe	1956/2009[a]
Deutsches Elektronen-Synchrotron (DESY)	Hamburg/Zeuthen	1959
Max Planck Institute for Plasma Physics (IPP)	Garching/Greifswald	1960
German Research Center for Environmental Health (KMGU)	München	1964
German Aerospace Center (DLR)	Köln	1969
GSI Center for Heavy Ion Research (GSI)	Darmstadt	1969
Center for Infection Research (HZI)	Braunschweig	1976
German Cancer Research Center (DKFZ)	Heidelberg	1976
Alfred Wegener Institute for Polar and Marine Research (AWI)	Bremerhaven/Potsdam/Sylt	1980
Center for Environmental Research (UFZ)	Leipzig/Halle/Magdeburg	1991
German Research Center for Geosciences (GFZ)	Potsdam	1992
Max Delbrück Center for Molecular Medicine (MDC)	Berlin-Buch	1992
German Center for Neurodegenerative Diseases (DZNE)	Bonn/Tübingen/Dresden	2009
Helmholtz Center Berlin for Materials and Energy (HZB)	Berlin	1957/2009[b]
Helmholtz Center Dresden-Rossendorf (HZDR)	Dresden	2011[c]

[a]The Forschungszentrum Karlsruhe (FZK), founded in 1956, merged with University of Karlsruhe in 2009 and formed the KIT.

[b]HZB is a merged entity of the former Hahn-Meitner Institute (founded in 1957) and the former Berlin Electron Storage Ring Company for Synchrotron Radiation (formerly a member of the Leibniz Association).

[b]HZDR is not a new entity, but was transferred from the Leibniz Association to the Helmholtz Association.

The majority of funding for the US National Laboratories comes from the DOE in the form of federal first-stream institutional core funding. Similarly, the Helmholtz centers have their core funding in institutional grants from the German Federal Government (90%) and from the respective Länder States wherein the labs reside (10%). However, institutional core funding is presently declining in both systems, which reflects both limitations of the financial capacities of the respective federal governments, and a political strategy in both countries to shift toward allocation of funding via soft money.[10]

Since the 1950s, the US National Laboratories have gone through three major growth-decline budget cycles, which have largely not correlated with numerical variations in the laboratory system. While the first budget expansion was directly connected to a steep growth in the total number of laboratories in the 1950s and the 1960s, the real terms budget decline of the 1970s occurred with no corresponding change in laboratory number. In contrast, the substantial budget increase in the 1980s coincided with only one newly founded lab. Budget austerity in the 1990s caused no laboratory shutdowns, and a return to budget growth in the early 2000s was not associated with any new laboratories. In comparison, the Helmholtz centers have experienced two similar major growth-decline budget cycles, which were also disconnected from the variation in total number of research centers. Substantial budget growth in the 1960s and 1970s was channeled into those laboratories founded in the 1950s. Furthermore, although the budget stagnated in the 1990s and 2000s, the number of laboratories increased.[11]

The first two US National Laboratories were founded on the remnants of the Manhattan Project, as a means to harness the weapons R&D resources for similar work in the postwar era, and expand them to other services for the military, the economy, and society at large.[12] Simultaneously, in 1947, three additional laboratories were created in other regions of the USA. The expansion period of this system of US National Laboratories lasted until the end of the 1960s, with particular growth occurring after the escalation of the Cold War in the mid- to late 1950s. By 1967, nine of the present ten civilian national laboratories had been established. In 1974, the steward agency of the labs, the Atomic Energy Commission (AEC), was replaced by the Energy Research and Development Administration (ERDA) as part of an attempt to better coordinate federal energy policy in the wake of the oil crisis.[13] This reform was also rooted in concerns that grew throughout the 1960s regarding the steeply increasing expenditure on National Laboratories, which were combined with a waning belief in the linear model of technological innovation, strong criticism toward the "military-industrial complex," and clear shifts in political priorities.[14]

The economic downturn in the 1970s caused some decline in spending in the US National Laboratories system, but this tendency again turned into growth with the renewed superpower competition and reinvigorated weapons programs spending in the 1980s. These changes also brought federal science spending in general to new heights, and launched several new projects. The previous economic downturn left lingering concerns over the role of Federal Laboratories in the national R&D system, which led to a series of legislative reforms in the 1980s, adding technology transfer and innovation to the laboratory missions.[15]

This trend continued throughout the 1990s. As the Cold War ended, the value of the spending on the National Laboratories came under severe criticism, leading to a rather dramatic downturn in laboratory funding. The Superconducting Super Collider project was closed before completion in 1993;[16] however, this case of termination has remained exceptional and did not create a precedent for any other US National Laboratory, despite their reduced budgets. In the 2000s, spending growth resumed and several major new projects were launched within the system, including the Spallation Neutron Source at Oak Ridge and the Linac Coherent Light Source (LCLS) at SLAC.

The first German national laboratories were founded in 1956, shortly after the lifting of the allied ban on nuclear research in the Federal Republic of Germany. A reactivation of the German research capabilities in nuclear physics had been promoted for several years by a strong lobbying group that stood ready to realize their plans once the ban was lifted. Between 1956 and 1959, no less than five large laboratories in the area of nuclear/PP were founded, and a designated Ministry for Atomic Matters was created. These efforts were accompanied by continuous reference to the emerging system of National Laboratories in the USA.[17] The following 15-year period witnessed significant expansions of the number of laboratories and the overall budget. Between 1964 and 1976, six new laboratories were founded and the overall inflation-adjusted budget almost tripled. These expansions included great diversification of the laboratories' research portfolios, from nuclear/PP to space and flight research, information technology, medicine, and biotechnology, among other areas.

Toward the end of the 1970s and into the 1980s, governmental authorities branded several nuclear research centers as having outlived their original purposes, and forced these centers to cut expenditures and personnel. As part of the same reevaluation of priorities, other laboratories were instructed to engage more actively in technology transfer and to diversify their activities for the benefit of society. Except for a short downturn in the early 1980s, the overall budget grew between 1977 and 1989. This growth included the founding of a new laboratory in polar and marine research (1983) and the launch of an additional funding stream toward the laboratories, which took the form of project funding schemes in thematically oriented areas and with specific funding opportunities for technology transfer activities.[18]

As a result of the 1990 German reunification, three new laboratories were founded in the early 1990s. In contrast to in the USA, the German system enjoyed institutional stability and even strength in the 1990s, largely due to the copying and extension of the governance structures of the formerly West German research organizations into the eastern part of the country. However, this expansion was not matched by any substantial funding increases, meaning that the new laboratories in the eastern part of Germany came at the expense of budget cuts suffered by the preexisting Western laboratories. In 2001, the Helmholtz Association was established as an umbrella organization within which all laboratories compete for individual shares of the overall five-year research budgets. While the German Federal Government remains the main sponsor, this reform made it less involved in agenda setting for the Helmholtz Centers and it has left most policy and decision making to the Helmholtz Association and its external peer reviewers.[19]

The key lesson drawn from these brief historical sketches is that the two laboratory systems have remained persistent and stable entities in their national public research systems, despite budgetary expansions and contractions and a series of substantial changes in their societal environments. This institutional stability sharply contrasts with the dramatic research portfolio changes that have occurred in all of these laboratories. At their founding in 1947, the original US laboratories had nuclear energy or nuclear energy-related R&D as original research mission. While the scope of this mission could be stretched quite far into several other areas of research, more or less at the discretion of lab directors, it was rather narrowly focused on nuclear energy in comparison with today's vast assortment of missions as regulated by the DOE and the US Congress: as chemical and molecular science, biological systems science, climate change science, applied materials science and engineering, and chemical engineering.[20] The original German laboratories were founded in the mid- to late 1950s as single-mission nuclear and PP centers, but today their research portfolios include climate change science, applied materials science and engineering, computer science, biotechnology, PS, astroparticle physics (APP), and chemical and molecular science.[21]

Given this vast expansion and change of the batteries of missions in the two systems of laboratories over more than 50 years, a central question is how such change has been accomplished on meso and micro levels, that is, on the level of laboratories and on the sublevel of research programs and large-scale infrastructure projects within the laboratories. To facilitate the analysis of change on these levels a conceptual framework will be introduced in Sect. 5.3 and put to use in the case analyses in Sect. 5.4.

5.3   PROCESSES OF GRADUAL ORGANIZATIONAL RENEWAL

Scholars in the study of institutional change have successfully developed two diametrically opposed versions of the concept of path dependence. On one hand, institutions can be sustained and reinforced through time by increasing returns and positive feedback processes. On the other hand, institutions can be formed at critical junctures provoked by radical change and the complementary identification of long periods of continuity and stability.[22] Recent advances in institutional theory complement these views, and argue that the processes and results of change should be considered variables in a theoretical framework that enables analysis of the gradual but cumulative adaptation of institutions.[23] The concept of incremental yet transformative change can also be applied to organizational change, and thus to the national laboratory systems of the USA and Germany, since they both seem to have evolved along gradual paths of organizational change rather than through events of radical system shocks.[24]

The fact that no laboratory in either of these two systems has ever been closed is testament to their institutional (macro level) persistence, as well as an indication that in general terms, the sponsorship relationships between the federal states and the laboratory systems have remained intact over time. Additionally, it appears that the overall major function of the laboratory systems is relatively stable within their respective national R&D systems. System (macro) level persistence might be viewed as an aggregation of continuity at the organizational (meso) level, meaning that the two national systems are stable because their constituent parts (the individual laboratories) are stable entities. This is true insofar as the laboratories are intact as organizational entities. However, as will be shown below, there exists considerable evidence of profound changes in the laboratory components (micro level), including the technical infrastructure, research fields, and organizational units. Therefore, it appears that gradual changes at the micro level have provided both the laboratories (meso level) and the two national systems (macro level) with the capacity to successfully adapt and survive over several decades.

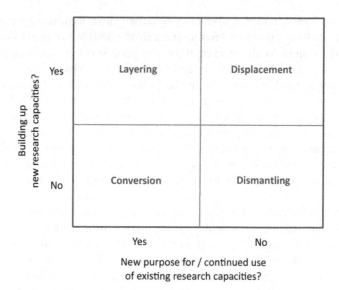

Fig. 5.1 Processes of gradual institutional change within research systems and research organizations

This conceptual scheme is designed such that the same analytical categories are applicable on all three levels (macro, meso, and micro). Figure 5.1 shows a cross-tabulation; the vertical axis indicates whether new research capacities are built up (including new technical infrastructure, the recruitment of scientists representing new research fields, or new organizational units), while the horizontal axis indicates whether existing research capacities continue to be used (including use for new purposes). The processes of gradual change in Fig. 5.1 are as follows. *Layering* is a process by which new arrangements are added on top of preexisting structures, thus enabling the accommodation of new elements without excessively compromising the logic of the preexisting structure. In contrast, *conversion* refers to when capacities for one set of goals are redirected to other ends, in a process that neither adds new capacities nor terminates the existing capacities. On the other hand, *displacement* means that research capacities are discontinued, as new ones are added in their place. Finally, *dismantling* simply means that research capacities—including technical infrastructures or research units—cease to be used without being replaced by new capacities.[25]

On the level of national laboratory systems, one straightforward process of gradual change is macro-level layering by the addition of new laboratories to the system. With few exceptions, this process occurs during concentrated time periods of expansion and diversification. Macro-level layering took place in the USA foremost in the 1950s and 1960s, and in Germany in the 1960s–1970s and during a short period in the early 1990s following German reunification. Outside of these periods, the two national systems have not grown numerically but only in terms of increasing budgets, which means that such budget growth has been absorbed by existing laboratories, thus indicating some form of micro-level layering (the addition of new research capacities), micro-level displacement (the substitution of existing research capacities for new, more expensive ones), or micro-level conversion (the redirection of existing capacities toward new, more expensive purposes and research fields), or any combination of these.

Micro-level changes can lead to meso-level transformations of whole laboratories. As will be shown below, DESY and SLAC are particularly interesting examples of how a series of intra-organizational (micro-level) changes can lead to full-scale organizational (meso-level) renewal. However, not all micro-level changes will necessarily cumulate into full-scale renewal at the laboratory level. The brief historical outlines in the previous section suggest that each federal government reevaluated their research policies and funding priorities in the wake of the economic downturn in the 1970s, and again at the end of the Cold War, which forced several laboratories to reconsider their missions and their planning.[26] However, while many laboratories initiated new projects and activities under the stewardship of their funders, these micro-level changes did not always lead to full-scale meso-level renewal with new dedicated research missions. Rather, several laboratories, especially when their budgets expanded, built on their multimission legacies and incorporated additional programs and projects into their portfolios without significantly altering their identities or mission statements but rather just increasing their diversification as an element in their pursued preservation of organizational status quo.[27] Therefore, while macro-level change is evident in the two systems, it is not simply linearly traceable back to micro-level changes—the accumulation of gradual changes inside labs into higher-level transformations is neither automatic nor straightforward.

Renewal can be examined in terms of three different dimensions: technical infrastructure, scientific fields, and organizational units. Change processes are typically multidimensional, multilevel, and multitemporal in the sense that a change process in one dimension, on one level, or on one timescale can translate to another change on another dimension, level, or timescale. For example, the layering of a new scientific activity with one piece of technical infrastructure on top of an existing one might eventually result in the new scientific activity taking over the piece of infrastructure. In this case, it would be possible to identify the layering of the new research field on top of the existing fields, then the dismantling of existing research areas, and finally the displacement of the original research field's use of the infrastructure by the new research field. Simultaneously, the components of the infrastructure itself might be layered, dismantled, replaced, or converted at various points in time and as part of the overall transformation.

In the next section, we will use the cases of DESY and SLAC to further analyze and exemplify this complex set of micro-level change processes that can lead to meso-level renewal. Thereafter, we will return to a discussion of how gradual changes on the micro level and renewal at the meso level relate to institutional persistence and stability of national systems of national laboratories on the macro level.

5.4   MULTILEVEL AND MULTIDIMENSIONAL RENEWAL AT DESY AND SLAC

Both DESY and SLAC were initially founded (in 1959 and 1961, respectively) as single-mission PP laboratories, each with one central piece of infrastructure. The construction and operation of these infrastructures was equal to the laboratory missions such that, in principle, both DESY and SLAC could have ceased to exist following the exhaustion of the scientific opportunities of these original machines.[28] As the laboratories continue to exist today, over 50 years later, we can conclude that this was not the actual course of events. Only a decade after their founding, each laboratory initiated construction projects for new major pieces of infrastructure for PP (see Fig. 5.2),[29] and continued to build several more PP machines for several decades. They also broadened their activities through the layering of a new research mission to operate machines for SR (or PS, as it was later called) on top of their original PP mission. This happened through several changes on the micro level, including changes in the overall scientific programs of the laboratories, in the uses of specific infrastructures and their technical setups and operations, and in the organizational units that were formally responsible for the scientific programs and infrastructures.

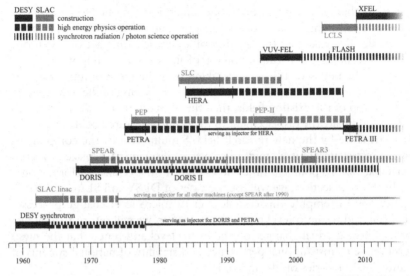

Fig. 5.2   Timeline of major infrastructures at DESY and SLAC, 1959–2015

Although the writing of the histories of these two labs with a one-sided focus on the infrastructures they have operated through the decades is oversimplified and would not give justice to the full range of micro-level processes that together bring about long-term change,[30] it is natural to use the succession of machines as a common thread in the analysis. The infrastructures form a key part of the missions of the labs and constitute powerful symbols of lab identities and culture, but most importantly, they are the key resources in the scientific programs of the laboratories. In the analysis below, clues regarding the combined gradual change processes at the micro level that cumulated into meso-level renewal of the two laboratories are therefore sought by focusing on the infrastructures—other publications use the necessary complementary perspectives.[31] Figures 5.3 and 5.4 illustrate the multilevel transformations of the two laboratories, highlighting their top level and overall 50-year changes (gray shading on the top level). We also point out some particularly evident examples of changes in infrastructure and science on the lower levels, which explain key component processes of the overall transformation (the gray-shaded ellipses lower in the figure). The gray-shaded ellipses should be interpreted as magnifications of those process elements shown with the same gray-shaded

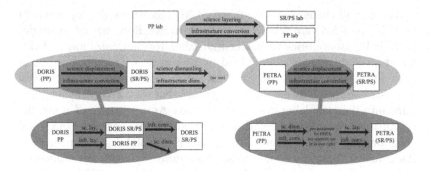

Fig. 5.3 Illustration of some key elements of the multilevel, long-term transformation of DESY with focus on infrastructures

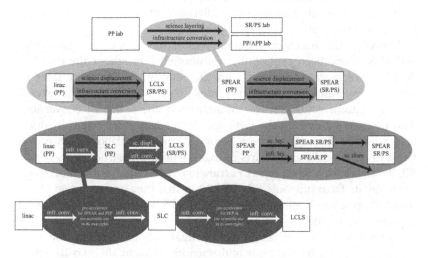

Fig. 5.4 Illustration of some key elements of the multilevel, long-term transformation of SLAC with focus on infrastructures

background on a higher level. They illustrate the increased level of detail that can be seen when analyzing change processes at a detailed level and with shorter time frames. In the second level of the figures, the arrows represent changes on the timescale of decades. On the third level, the arrows represent change processes that typically take a few years.

The overall changes of DESY and SLAC (shown by the top levels of Figs. 5.3 and 5.4) are relatively straightforward. Both started as single-mission PP laboratories with a central laboratory organization and different auxiliary activities conducted by user groups. Organizationally, at both DESY and SLAC, the early SR research comprised of peripheral activities conducted by external user groups that were eventually incorporated into the main laboratory organizations. As organizational units, the synchrotron radiation labs within DESY and SLAC, named HASYLAB (Hamburger Synchrotronstrahlungslabor, Hamburg Synchrotron Radiation Laboratory) and SSRL (Stanford Synchrotron Radiation Laboratory), were founded as distinct entities in the late 1970s, and they became organizational divisions of SLAC (1990s) and DESY (2000s), respectively. Today, DESY still includes a PP division, and SLAC includes a combined APP and PP division. Thus, the 50-year histories of both DESY and SLAC as organizations can be summarized as the addition of SR/PS as a new research mission, which diversifies the former single-mission laboratories (laboratory level: *science layering*).

However, the underlying assumption of this chapter is that DESY and SLAC have been profoundly transformed throughout the past five decades, not merely expanded with the addition of one more layer of activities over an unchanged core mission. We argue that the overall 50-year transformation on the infrastructure side is one of conversion. This premise is based on the facts that both laboratories originally operated scientific infrastructure solely for PP, and both laboratories modified and rebuilt substantial parts of that scientific infrastructure to enable SR/PS (laboratory level: *infrastructure conversion*), and both laboratories are today de facto primarily SR/PS labs in that they operate some of the world's top research infrastructures for SR/PS while not running any PP experiments/machines.

Compared to the analysis of formal organizational changes, the analyses of research infrastructures and scientific fields at the two laboratories are significantly more complex. It must be acknowledged that the organizational changes are unthinkable without the preceding changes to major technical installations and the science around them. The laboratory histories clearly show that a delay preceded their organizational transformations, that is, SR/PS received two formal organizational units/divisions only some time after the scientific–technical change had occurred. It is also important to note that the formal organizational SR/PS units did not replace existing ones, but were instead added on top of existing organizational structures. Due to these factors, the following detailed analysis portrays the organizational side as somewhat less prominent than the other two dimensions (*infrastructure* and *science*), but this is due to a deliberate choice of perspective and emphasis in this chapter.

Figure 5.3 details some key changes to the infrastructure and science of DESY. The science layering and infrastructure conversion at the top level are disaggregated into the analyses of the transformations of key research infrastructures DORIS (Doppel-Ring Speicher, Double Storage Ring) and PETRA (Positron-Elektron Tandem Ringanlage, Positron-Electron Tandem Ring Facility) from DORIS (PP) to DORIS III (SR/PS) and from PETRA (PP) to PETRA III (SR/PS).[32] Both transformations are characterized as processes of simultaneous *science displacement* and *infrastructure conversion* (second level in Fig. 5.3), and then further disaggregated at the machine level (third level in Fig. 5.3). DORIS was originally built as a storage ring for PP, with construction beginning in 1968. Between 1974 and 1992, DORIS was additionally used in parallel for SR in so-called parasitic mode,[33] which required some additional instrumentation (*science layering* and *infrastructure layering*). In 1993, the PP program at DORIS was canceled and the machine became fully dedicated to SR, which means it underwent final *infrastructure conversion* and *science dismantling* (of PP).

PETRA is an even larger storage ring for PP, for which construction began in 1975. In 1986, PETRA was closed for scientific use and turned into a pre-accelerator for the much larger HERA (Hadron-Elektron Ringanlage, Hadron Electron Ring Facility), run until 2007 (*science dismantling* and *infrastructure conversion*). Later, PETRA was turned into a SR source (PETRA III, *science layering* and *infrastructure conversion*), which eventually, in 2012, made DORIS redundant as a SR facility. At the level of technical infrastructure, the construction of ever-larger machines at DESY over a 50-year time frame can be interpreted as a multistep process of *infrastructure layering* (addition of new machines for PP) and of *infrastructure conversion* (using smaller synchrotrons as injectors for larger storage rings, and dedicating old storage rings to PS) (Fig. 5.2).

Similarly, Fig. 5.4 details some key changes to the infrastructure and science of SLAC. The science layering and infrastructure conversion at the top level are disaggregated into the analyses of the transformations of the key research infrastructures the SLAC original linac and the SPEAR (Stanford Positron-Electron Accelerator Ring) machine, from linac (PP) to LCLS (SR/PS), and from SPEAR (PP) to SPEAR (SR/PS). Both transformations are characterized as processes of simultaneous *science displacement* and *infrastructure conversion* (second level in Fig. 5.4), and then further disaggregated at the machine level (third and fourth levels in Fig. 5.4). The SLAC linac was originally built for PP, but in 1972 it was converted for use merely as a pre-accelerator for other SLAC machines (*infrastructure conversion*). Then, in the 1980s, the linac was used to construct the all-particle physics SLC (SLAC Linear Collider) machine (*infrastructure conversion*). After the SLC closed in the late 1990s, two-thirds of the linac was used as a pre-accelerator for PEP-II (Positron-Electron Project), thus once again undergoing infrastructure conversion, and later, the other one-third was used as a key piece in the construction of the LCLS, which is a state-of-the-art free electron laser machine for PS (yet another instance of *infrastructure conversion*). The LCLS opened for scientific use in 2009. The several-step *infrastructure conversion* from the original 1960s linac to the 2000s LCLS also represents a process of long-term *science displacement* since a key piece of infrastructure previously used solely for PP is now used solely for PS.

SPEAR is a storage ring that was designed and built for use in PP, starting in 1970. The scientific use of SPEAR was soon extended to include a SR program, which required some additional instrumentation (*science layering* and *infrastructure layering*). By the early 1990s, PP research at SPEAR was cancelled in favor of the SR program, which completely took over operations at SPEAR (*science dismantling*). At the level of technical infrastructure, the construction of ever-larger machines at SLAC over a 50-year time frame can be interpreted as a multistep process of *infrastructure layering* (addition of new machines for PP) and of *infrastructure conversion* (using the original linac as an injector for larger machines and dedicating the old storage ring SPEAR to PS) (Fig. 5.2).

We have disaggregated the cases of DESY and SLAC in some detail, in order to exemplify an analysis of micro-level change processes that led to meso-level renewal. The comparison of the two laboratories reveals striking similarities. Both laboratories initiated the construction of storage rings for PP (DORIS and SPEAR) approximately ten years after their founding, which later turned out to be extremely useful for SR research. Viewed from today, when neither one of them is in use for PP anymore, the overall transformation of these storage rings for PP comprises *infrastructure conversion* and *science displacement* (second level, to the right, in Figs. 5.3 and 5.4). On more detailed level, the transformations of DORIS and SPEAR occurred through a gradual addition of SR activities (and associated instrumentation) to the machines (*science layering* and *infrastructure layering*; third level, on the right side, in Figs. 5.3 and 5.4). This was followed by abandonment of the DORIS and SPEAR rings by PP (*science dismantling*), and the concurrent adaptation of the machines for optimized SR operation (*infrastructure conversion*). DORIS was later shut down (*infrastructure dismantling* and *science dismantling*) in 2012, while SPEAR remains in operation, serving the SR user community.

In the late 1970s, both DESY and SLAC built larger storage rings for PP. The SLAC storage ring PEP was almost exclusively used for PP, with only some sporadic SR operations undertaken in the 1980s. PEP was eventually converted into PEP-II and taken out of operation in 2008 (this development is not shown in Fig. 5.4). At DESY, the PETRA storage ring was used solely for PP research for several years, and was then turned into a pre-accelerator for the much larger HERA particle physics machine (*science dismantling* and *infrastructure conversion*; third level, to the left, in Fig. 5.3). Upon the closing of HERA in 2007 (this development is not shown in Fig. 5.3), PETRA was rebuilt into a SR facility (*infrastructure conversion* and *science layering*; third level, on the left side, in Fig. 5.3) and has been used for this purpose since 2009.

The parallels between the changes at DESY and SLAC are further underscored when the machines are displayed on the same timeline (Fig. 5.2). As previously mentioned, at the level of technical infrastructure, the construction of ever-larger machines at both DESY and SLAC over a 50-year time frame can be interpreted as a multistep process of *infrastructure layering* (addition of new machines for PP) and *infrastructure conversion* (using the original machines as injectors for larger machines and dedicating sold storage rings to PS). In each case, this succession culminates in the construction of new infrastructure designed for and dedicated to PS. At DESY, this is the construction and operation of the VUV-FEL (Vacuum-Ultraviolet Free Electron Laser, later renamed FLASH, Free Electron Laser Hamburg) in the late 1990s, and the start of construction of XFEL (X-ray Free Electron Laser) in 2009.[34] At SLAC, this is the 2003–2009 construction of LCLS, which uses parts of the original SLAC linac and thus represents an infrastructure conversion.

While we observe several cases of infrastructure conversion paired with layering of new scientific fields, there are also examples of infrastructure changes that were not combined with respective changes in science. HERA (at DESY) and PEP-II (at SLAC) are examples of dismantling of technical infrastructure that meant science dismantling (of PP activities) on the level of the machines but, importantly, not on the level of the labs. With no future use in sight, HERA was shut down in 2007 and PEP-II in 2008. However, large data sets from experiments at these two machines remained to be analyzed, and thus many particle physicists remained at the two laboratories to complete this work.

All new machines designed and built at DESY and SLAC before the mid-1990s started out as dedicated PP facilities, and all have either been gradually converted into SR facilities (DORIS and SPEAR gradually, PETRA recently and comparably abruptly) or dismantled (PEP and HERA), or both (DORIS). Thus, while we observe several major instances of PP displacement and dismantling at the level of the machines, there has been no equivalent displacement of PP at the laboratory level (yet). PP remains part of their stated core missions, though it is now a somewhat less prominent scientific field. The fact that PP was not immediately dismantled upon closure of the technical infrastructure of this research field shows that scientific programs are only partly tied to infrastructures—appearing to even function independently of them to some extent. This also explains why micro-level change processes in one dimension (e.g., infrastructure) are not necessarily identical to change processes in another dimension (e.g., scientific fields). Of course, one key question is how long the scientific programs of PP can continue without operating a machine. Somewhat speculatively, the material at hand and the analysis above point in the direction of a full eventual displacement of PP, partly by APP and most importantly by SR/PS, seen in long-term and laboratory-level perspective, at both labs.

5.5   CONCLUSION

This chapter addresses the question of why none of the national laboratories of Germany and the USA have ever been closed, despite considerable changes or even the decline and expiration of their original research missions. The analysis has shown that the answer to this question is complex, since research laboratory renewal is a multilevel and a multitemporal process. We propose that analysis of the complexity of research organizations and their changes requires data spanning several decades, and observations in (at least) three dimensions: technical infrastructures, scientific fields, and organizational units. The combination of these three dimensions within and across certain time windows is necessary to unveil and understand the organizational process of change. Our present analysis touches upon several possible answers, some of which we believe are worthy of more attention in future research.

First, we argue that organizational renewal involves gradual changes at the micro level, which typically do not threaten the existing routines and capacities of research laboratories with regard to technical infrastructure and scientific fields. However, gradual changes can complement each other and, through mutual cumulation over extended periods of time, can lead to reorientations of entire laboratories that go far beyond the short-term small-scale developments. Thus, gradual but cumulative processes of change can have discontinuous effects on the scientific missions of laboratories and their respective research capacities. This link is particularly visible at DESY and SLAC, where we observe a major shift from PP research to PS (although PP remains, and APP has also been added). Although we have not discussed in this chapter what caused these micro-level change processes to occur and then to cumulate, we know from the histories of the two labs that institutional entrepreneurs, laboratory leadership, universities in the vicinity of national laboratories, and federal sponsorship were key elements in explaining how micro-level investments in exploration cumulate into meso-level renewal. Further empirical research is needed to generalize these findings.

Second, the translation of micro-level changes into meso-level renewal is neither automatic nor straightforward, but rather a complex multilevel and multitemporal process. Thus, we would require more knowledge about "failed" laboratories, that is, facilities that have not successfully adapted to changing societal, economic, and political circumstances. In the 1970s, during the consolidation phase of the national laboratories system in the USA, the federal government organized a series of reviews. The aim was to determine whether any research programs within the national laboratories required adaptation, or if perhaps entire laboratories should be closed, as part of the government downsizing promised by the Reagan administration. Silicon Valley entrepreneur David Packard headed one of these review panels, and reportedly "chilled the hearts of laboratory directors across the nation"[35] by saying "Preservation of the laboratory is not a mission."[36] This statement is clearly provocative, but there is little empirical evidence to substantiate it. As no laboratories were ultimately closed, the question remains under which institutional conditions laboratories fail to translate micro-level changes into meso-level renewal, and what consequences this has on their scientific productivity and impact.

Third, we have argued that macro-level stability is related to both micro-level and meso-level changes within and across single laboratories. Our present analyses provide no conclusive evidence demonstrating which level is more important in this regard. The transformations of DESY and SLAC evidently support the claim that successful adaptations at the meso level tend to stabilize the laboratory systems as a whole. However, since DESY and SLAC each represent only one laboratory in their respective national systems, we cannot generalize this statement without providing supportive empirical evidence relating to the other 16 German and 9 US laboratories. Still, we know that micro-level changes have occurred in one way or another in all national laboratories in these two countries. As mentioned above, since the consolidation phase of each national laboratory system, their budget growth has typically been consumed by existing laboratories but not by new ones. Additionally, the original national laboratories had core research missions of nuclear energy or nuclear energy-related R&D, but their research portfolios later broadened considerably into areas including chemical and molecular science, biological systems science, climate change science, applied materials science and engineering, chemical engineering, computer science, biotechnology, and APP. Therefore, it seems that micro-level changes in single national laboratories have provided the macro-level system with enough adaptive capacity to survive despite considerable macro-level changes in research policy and society at large, such as those brought on by the end of the Cold War.

The explanation of how micro-level adaptation and meso-level renewal influence macro-level stability or change, and vice versa, is key to understanding institutional change in national laboratory systems. One possibility is that micro-level changes in single national laboratories have provided the macro level with enough adaptive capacity to maintain its status quo (i.e., the survival of all national laboratories ever founded). The outcome of this situation would be very different compared to a situation where micro-level changes cumulate into meso-level renewal, thus providing the macro level with renewability and survival capacity. System level reproduction by micro-level adaptation is quite different from system level transformation by meso-level renewal. Although we know that the two national laboratory systems have survivor qualities, we do not yet know whether the renewal of DESY and SLAC can be generalized to other national laboratories. This challenging question remains on the agenda for future research.

NOTES

1. Daniel S. Greenberg, *The Politics of Pure Science*, 2nd ed. (Chicago, IL: University of Chicago Press, 1999/1967); Peter J. Westwick, *The National Laboratories: Science in an American System 1947–1974* (Harvard University Press, 2003); Bruce L. R. Smith, *American Science Policy since World War II* (Washington, DC: Brookings Institute, 1990).

2. For the US case, we delimit the analysis to those National Laboratories that have civilian missions and civilian oversight in the shape of the Office of Science of the United States Department of Energy (DOE). The seven weapons laboratories are excluded from the analysis since their activities and organizations are classified (secret) and thus do not lend themselves to this type of study. For obvious historical reasons, a similar delimitation is not necessary in the German case, where no weapons laboratories of the same sort exist.

3. Olof Hallonsten, "The parasites: Synchrotron radiation at SLAC, 1972–1992," *Historical Studies in the Natural Sciences* 45, no. 2 (2015); Thomas Heinze, Olof Hallonsten, and Steffi Heinecke, "From Periphery to Center. Synchrotron radiation at DESY, Part I: 1962–1977," *Historical Studies in the Natural Sciences* 45, no. 3 (2015a); Thomas Heinze, Olof Hallonsten, and Steffi Heinecke, "From Periphery to Center. Synchrotron radiation at DESY, Part II: 1977–1993," *Historical Studies in the Natural Sciences* 45, no. 4 (2015b).

4. Robert P. Crease, "The National Synchrotron Light Source, Part I: Bright Idea," *Physics in perspective* 10 (2008); Robert P. Crease, "The National Synchrotron Light Source, Part II: The Bakeout," *Physics in perspective* 11 (2009); Park Doing, *Velvet Revolution at the Synchrotron: Biology, Physics, and Change in Science* (Cambridge, MA: MIT Press, 2009); Olof Hallonsten, "Growing Big Science in a Small Country: MAX-lab and the Swedish Research Policy System," *Historical Studies in the Natural Sciences* 41, no. 2 (2011); Olof Hallonsten and Thomas Heinze, "Formation and Expansion of a New Organizational Field in Experimental Science," *Science and Public Policy* 42, no. 6 (2015); Catherine Westfall, "Retooling for the Future: Launching the Advanced Light Source at Lawrence's Laboratory, 1980–1986," *Historical Studies in the Natural Sciences* 38, no. 4 (2008b); Catherine Westfall, "Institutional Persistence and the Material Transformation of the US National Labs: the Curious Story of the Advent of the Advanced Photon Source," *Science and Public Policy* 39, no. 4 (2012).

5. The chapter partly draws on material published in Olof Hallonsten and Thomas Heinze, "Institutional persistence through gradual adaptation: Analysis of national laboratories in the USA and Germany," *Science and Public Policy* 39 (2012): 436; and Olof Hallonsten and Thomas Heinze,

"From particle physics to photon science: Multidimensional and multilevel renewal at DESY and SLAC," *Science and Public Policy* 40, no. 5 (2013).

6. Before 1974, the Atomic Energy Commission (AEC); between 1974 and 1977, the Energy Research and Development Administration (ERDA); since 1977, the DOE.

7. Westwick, *National Laboratories;* Michael Crow and Barry Bozeman, *Limited by Design. R&D Laboratories in the U.S. National Innovation System* (New York: Columbia University Press, 1998).

8. Hans-Willy Hohn and Uwe Schimank, *Konflikte und Gleichgewichte im Forschungssystem: Akteurkonstellationen und Entwicklungspfade in der staatlich finanzierten außeruniversitären Forschung* (Frankfurt, Germany: Campus, 1990), 39–62; Thomas Heinze "Trends und Entwicklungslinien der außeruniversitären Forschung im internationalen Vergleich," in *Wissenschaft als Beruf. Bestandsaufnahme—Diagnosen—Empfehlungen,* ed. M. Haller (Wien: Österreichische Akademie der Wissenschaften, 2013).

9. Thomas Heinze and Natalie Arnold, "Governanceregimes im Wandel. Eine Analyse des außeruniversitären, staatlich finanzierten Forschungssektors in Deutschland," *Kölner Zeitschrift für Soziologie und Sozialpsychologie* 60 (2008).

10. Hallonsten and Heinze, "Institutional persistence," 436.

11. Ibid., 452–53.

12. Richard G. Hewlett and Oscar E. Anderson, Jr., *A History of the United States Atomic Energy Commission. Volume 1. The New World, 1939/1946* (University Park, PA: Pennsylvania State University Press, 1962), 714–22; Terrence R. Fehner and Jack M. Holl, *Department of Energy 1977–1994. A Summary History* (Washington, DC: United States Department of Energy, 1994), 11.

13. Jack M. Holl, *Argonne National Laboratory 1946–96* (Chicago, IL: University of Illinois Press, 1997), 279; Catherine Westfall, "Surviving the squeeze: National laboratories in the 1970s and 1980s," *Historical Studies in the Natural Sciences* 38 (2008a): 476.

14. Hallam Stevens, "Fundamental physics and its justifications, 1945–1993," *Historical Studies in the Physical and Biological Sciences* 34 (2003): 161; Westwick, *National Laboratories,* 269.

15. David H. Guston, "Stabilizing the boundary between US politics and science: The role of the Office of Technology Transfer as a boundary organization," *Social Studies of Science* 29 (1999): 94; Ann Johnson, "The end of pure science: Science policy from Bayh–Dole to the NNI," in *Discovering the Nanoscale,* ed. D. Baird et al. (Amsterdam, the Netherlands: IOS Press, 2004); Westfall, "Surviving the squeeze."

16. Michael Riordan, Lillian Hoddeson, and Adrienne W. Kolb, *Tunnel Visions. The Rise and Fall of the Superconducting Super Collider* (Chicago, IL; University of Chicago Press, 2016), p. 201–247.

17. Gerhard A. Ritter, *Großforschung und Staat in Deutschland. Ein historischer Überblick* (München, Germany: Beck, 1995), 56–77.
18. Hohn and Schimank, *Konflikte und Gleichgewichte*, 262–95; Ritter, *Großforschung und Staat*, 100–11.
19. Heinze and Arnold, "Governanceregimes"; Sabine Helling-Moegen, *Forschen nach Programm. Die programmorientierte Förderung in der Helmholtz-Gemeinschaft: Anatomie einer Reform* (Marburg: Tectum Verlag, 2009); Insa Pruisken, *Fusionen im institutionellen Feld 'Hochschule und Wissenschaft'* (Baden-Baden: Nomos, 2014), 157–81.
20. Robert P. Crease, *Making Physics: A Biography of Brookhaven National Laboratory, 1946–1972* (Chicago: The University of Chicago Press, 1999); Holl, *Argonne National Laboratory*; Leland Johnson and Daniel Schaffer, *Oak Ridge National Laboratory: The First Fifty Years* (Memphis, TN: The University of Tennessee Press, 1994); Westwick, *National Laboratories*.
21. Ritter, *Großforschung und Staat*; Claus Habfast, *Großforschung Mit Kleinen Teilchen. Das Deutsche Elektronen-Synchrotron Desy 1956–1970* (Heidelberg: Springer Verlag, 1989).
22. James Mahoney, "Path Dependence in Historical Sociology," *Theory and Society* 29 (2000); Paul Pierson, *Politics in Time: History, Institutions, and Political Analysis* (Princeton, NJ: Princeton University Press, 2004).
23. Kathleen Thelen, "How institutions evolve. Insights from comparative historical analysis," in *Comparative Historical Analysis in the Social Sciences*, ed. J. Mahoney, and D. Rueschemeyer (New York: Cambridge University Press, 2003); Wolfgang Streeck and Kathleen Thelen, "Introduction: Institutional change in advanced political economies," in *Beyond Continuity. Institutional Change in Advanced Political Economies*, ed. W. Streeck, and K. Thelen (Oxford, UK: OUP, 2005); Wolfgang Streeck, *Re-Forming Capitalism. Institutional Change in the German Political Economy* (Oxford, UK: OUP, 2009).
24. Hallonsten and Heinze, "Institutional persistence."
25. Figure 5.1 is an adapted version of the one published in Hallonsten and Heinze, "From particle physics to photon science."
26. See, for example, Westfall, "Surviving the squeeze."; Stevens, "Fundamental physics," 183–96; Daniel Kevles, "Big Science and big politics in the United States: Reflections on the death of the SSC and the life of the Human Genome Project," *Historical studies in the physical and biological sciences* 27, no. 2 (1997).
27. Holl, *Argonne National Laboratory*, 46; Westfall, "Surviving the squeeze"; Helling-Moegen, *Forschen nach Programm*; Pruisken, *Fusionen*, 157–81.
28. Wolfgang K.H. Panofsky, *Panofsky on Physics, Politics, and Peace: Pief Remembers* (New York: Springer, 2007), 126; Erich Lohrmann and Paul

Söding, *Von schnellen Teilchen und hellem Licht: 50 Jahre Deutsches Elektronen-Synchrotron DESY,* 2nd edition (Berlin: Wiley, 2013), 1–19.
29. Figure 5.2 shows a combined and updated version of two figures published in Hallonsten and Heinze, "From particle physics to photon science," 594, 597.
30. Hallonsten, "The parasites."
31. Hallonsten, "The parasites"; Heinze, Hallonsten, and Heinecke, "DESY, Part I."; Heinze, Hallonsten, and Heinecke, "DESY, Part II."
32. Note that the infrastructure conversion of upgrading individual machines often coincided with name changes, such as the upgrade from DORIS to DORIS II. For reasons of clarity, we use only the main names and not the names of individual versions in Figs. 5.3 and 5.4.
33. For details on this term, see Hallonsten, "The parasites," 219.
34. Although, for the sake of stringency and correctness, it should be noted that the XFEL is not part of DESY but a stand-alone company owned by 12 European governments through their respective research councils and ministries. The German share of 58% of the company is administered by DESY which acts as the German government's representative in the XFEL governing bodies, and part of the XFEL is also physically located to the DESY site, which means that the XFEL in practice (though not legally) is partly integrated into the DESY organization. Olof Hallonsten, "The Politics of European Collaboration in Big Science," in *The Global Politics of Science and Technology—Vol. 2, Global Power Shift,* ed. M. Mayer, M. Carpes, and R. Knoblich (Berlin: Springer, 2014).
35. Westfall, "Retooling for the Future," 571.
36. Holl, *Argonne National Laboratory,* 401.

Acknowledgments This paper is based on research that was supported by the German Federal Ministry for Education and Research (BMBF) grant 01UZ1001. We are grateful for helpful comments and suggestions from Arlette Jappe, Thomas Kaiserfeld, and Richard Münch, and participants at the "International Conference on Intellectual and Institutional Innovation in Science" at the Berlin-Brandenburg Academy of Sciences and Humanities, September 13–15, 2012, and the workshop "The New Big Science" at Lund University, January 16–17, 2013, where we presented early versions of this paper.

REFERENCES

Crease, Robert P. 1999. *Making physics: A biography of Brookhaven National Laboratory, 1946–1972*. Chicago: The University of Chicago Press.

Crease, Robert P. 2008. The National Synchrotron Light Source, part I: Bright idea. *Physics in Perspective* 10: 438–467.

Crease, Robert P. 2009. The National Synchrotron Light Source, part II: The bakeout. *Physics in Perspective* 11: 15–45.

Crow, Michael, and Barry Bozeman. 1998. *Limited by design. R&D laboratories in the U.S. national innovation system*. New York: Columbia University Press.

Doing, Park. 2009. *Velvet revolution at the synchrotron: Biology, physics, and change in science*. Cambridge, MA: MIT Press.

Fehner, Terrence R., and Jack M. Holl. 1994. *Department of Energy 1977–1994: A summary history*. Washington, DC: United States Department of Energy.

Greenberg, Daniel S. 1999/1967. *The politics of pure science*, 2nd ed. Chicago: University of Chicago Press.

Guston, David H. 1999. Stabilizing the boundary between US politics and science: The role of the Office of Technology Transfer as a boundary organization. *Social Studies of Science* 29: 87–111.

Habfast, Claus. 1989. *Großforschung Mit Kleinen Teilchen. Das Deutsche Elektronen-Synchrotron Desy 1956–1970*. Heidelberg: Springer.

Hallonsten, Olof. 2011. Growing big science in a small country: MAX-lab and the Swedish research policy system. *Historical Studies in the Natural Sciences* 41(2): 179–215.

Hallonsten, Olof. 2014. The politics of European collaboration in big science. In *The global politics of science and technology*, vol. 2, ed. M. Mayer, M. Carpes, and R. Knoblich, 31–46. Berlin: Springer.

Hallonsten, Olof. 2015. The parasites: Synchrotron radiation at SLAC, 1972–1992. *Historical Studies in the Natural Sciences* 45(2): 217–272.

Hallonsten, Olof, and Thomas Heinze. 2012. Institutional persistence through gradual adaptation: Analysis of national laboratories in the USA and Germany. *Science and Public Policy* 39: 450–463.

Hallonsten, Olof, and Thomas Heinze. 2013. From particle physics to photon science: Multidimensional and multilevel renewal at DESY and SLAC. *Science and Public Policy* 40(5): 591–603.

Hallonsten, Olof, and Thomas Heinze. 2015. Formation and expansion of a new organizational field in experimental science. *Science and Public Policy* 42(6): 841–854.

Heinze, Thomas. 2013. Trends und Entwicklungslinien der außeruniversitären Forschung im internationalen Vergleich. In *Wissenschaft als Beruf. Bestandsaufnahme—Diagnosen—Empfehlungen*, ed. M. Haller, 74–87. Wien: Österreichische Akademie der Wissenschaften.

Heinze, Thomas, and Natalie Arnold. 2008. Governanceregimes im Wandel. Eine Analyse des außeruniversitären, staatlich finanzierten Forschungssektors in Deutschland. *Kölner Zeitschrift für Soziologie und Sozialpsychologie* 60: 686–722.

Heinze, Thomas, and Richard Münch. 2012. Intellektuelle Erneuerung der Forschung durch institutionellen Wandel. In *Institutionelle Erneuerungsfähigkeit der Forschung*, ed. T. Heinze and G. Krücken, 15–38. Wiesbaden: VS Verlag.

Heinze, Thomas, Olof Hallonsten, and Steffi Heinecke. 2015a. From Periphery to Center. Synchrotron radiation at DESY, part I: 1962–1977. *Historical Studies in the Natural Sciences* 45(3): 447–492.

Heinze, Thomas, Olof Hallonsten, and Steffi Heinecke. 2015b. From Periphery to Center. Synchrotron radiation at DESY, part II: 1977–1993. *Historical Studies in the Natural Sciences* 45(4): 513–548.

Helling-Moegen, Sabine. 2009. *Forschen nach Programm. Die programmorientierte Förderung in der Helmholtz-Gemeinschaft: Anatomie einer Reform.* Marburg: Tectum Verlag.

Hewlett, Richard G., and Oscar E. Anderson Jr. 1962. *A history of the United States Atomic Energy Commission. Volume 1. The new world, 1939/1946.* University Park: Pennsylvania State University Press.

Hohn, Hans-Willy, and Uwe Schimank. 1990. *Konflikte und Gleichgewichte im Forschungssystem: Akteurkonstellationen und Entwicklungspfade in der staatlich finanzierten außeruniversitären Forschung.* Frankfurt: Campus.

Holl, Jack M. 1997. *Argonne National Laboratory 1946–96.* Chicago: University of Illinois Press.

Johnson, Ann. 2004. The end of pure science: Science policy from Bayh–Dole to the NNI. In *Discovering the nanoscale*, ed. D. Baird, A. Nordmann, and J. Schummer, 217–230. Amsterdam: IOS Press.

Johnson, Leland, and Daniel Schaffer. 1994. *Oak Ridge National Laboratory: The first fifty years.* Memphis: The University of Tennessee Press.

Kevles, Daniel. 1997. Big science and big politics in the United States: Reflections on the death of the SSC and the life of the Human Genome Project. *Historical Studies in the Physical and Biological Sciences* 27(2): 269–297.

Lohrmann, Erich, and Paul Söding. 2013. *Von schnellen Teilchen und hellem Licht: 50 Jahre Deutsches Elektronen-Synchrotron DESY*, 2nd ed. Berlin: Wiley.

Mahoney, James. 2000. Path dependence in historical sociology. *Theory and Society* 29: 507–548.

Panofsky, Wolfgang K.H. 2007. *Panofsky on physics, politics, and peace: Pief remembers.* New York: Springer.

Pierson, Paul. 2004. *Politics in time: History, institutions, and political analysis.* Princeton: Princeton University Press.

Pruisken, Insa. 2014. *Fusionen im institutionellen Feld 'Hochschule und Wissenschaft'.* Baden-Baden: Nomos.

Riordan, M., Hoddeson, L. and Adrienne W. Kolb. (ed.) 2016. *Tunnel Visions. The Rise and Fall of the Superconducting Super Collider*. Chicago: University of Chicago Press.

Ritter, Gerhard A. 1995. *Großforschung und Staat in Deutschland. Ein historischer Überblick*. München: Beck.

Smith, Bruce L.R. 1990. *American science policy since World War II*. Washington, DC: Brookings Institute.

Stevens, Hallam. 2003. Fundamental physics and its justifications, 1945–1993. *Historical Studies in the Physical and Biological Sciences* 34: 151–197.

Streeck, Wolfgang. 2009. *Re-forming capitalism. Institutional change in the German political economy*. Oxford: Oxford University Press.

Streeck, Wolfgang, and Kathleen Thelen. 2005. Introduction: Institutional change in advanced political economies. In *Beyond continuity. Institutional change in advanced political economies*, ed. W. Streeck and K. Thelen, 1–39. Oxford: Oxford University Press.

Thelen, Kathleen. 2003. How institutions evolve. Insights from comparative historical analysis. In *Comparative historical analysis in the social sciences*, ed. J. Mahoney and D. Rueschemeyer, 208–240. New York: Cambridge University Press.

Westfall, Catherine. 2008a. Surviving the squeeze: National Laboratories in the 1970s and 1980s. *Historical Studies in the Natural Sciences* 38: 475–478.

Westfall, Catherine. 2008b. Retooling for the future: Launching the advanced light source at Lawrence's Laboratory, 1980–1986. *Historical Studies in the Natural Sciences* 38(4): 569–609.

Westfall, Catherine. 2012. Institutional persistence and the material transformation of the US National Labs: The curious story of the advent of the Advanced Photon Source. *Science and Public Policy* 39(4): 439–449.

Westwick, Peter J. 2003. *The National Laboratories: Science in an American system 1947–1974*. Cambridge: Harvard University Press.

CHAPTER 6

Institutional Context and Growth of New Research Fields. Comparison Between State Universities in Germany and the USA

Arlette Jappe and Thomas Heinze

6.1 INTRODUCTION

This chapter examines how universities build up and expand research capacities in new and emerging scientific fields following major scientific breakthroughs. The research question is to what extent the institutional framework in which universities are embedded supports such expansion and renewal. Scientific research is oriented toward two opposing values: innovation and tradition.[1] Research thus is characterized by a fundamental tension between forces that on the one hand attempt to leave conventional paths of thought and transcend established doctrines and on the other hand seek conformity to disciplinary research and accepted frameworks. James March introduced the terms *exploration* and *exploitation* to describe this fundamental tension.[2] Exploration designates the search

A. Jappe (✉) • T. Heinze
Interdisciplinary Centre for Science and Technology Studies (IZWT),
University of Wuppertal, Wuppertal, Germany

© The Editor(s) (if applicable) and The Author(s) 2016
T. Heinze, R. Münch (eds.), *Innovation in Science and Organizational Renewal*, DOI 10.1057/978-1-137-59420-4_6

for new knowledge and overcoming of current know-how, and exploitation designates the refinement and validation of established knowledge and incorporation of new findings into existing patterns of thought. Exploration opens up new horizons and perspectives while exploitation enhances existing knowledge and technology (see editor's introduction).

The tension between *exploration* and *exploitation* can be investigated from two angles. First, we may ask which institutional conditions facilitate the emergence of research breakthroughs. From this perspective, institutional conditions for the emergence of new scientific solutions are investigated.[3] Second, we can also inquire into conditions for the propagation and diffusion of scientific inventions. If something new has been invented, how does it take hold over existing approaches? How are innovators able to overcome both the inertia and resistance of the scientific establishment? As far as the analysis of renewal in science and technology is concerned, the second perspective seems more relevant. Therefore, in this article, we investigate the capability of universities to seize upon and expand new and innovative research fields.

To do so, we chose two research breakthroughs from the recent past with an impact that can be adequately investigated from a sociological point of view: the scanning tunneling microscope (STM), developed in 1982 by Gerd Binnig and Heinrich Rohrer at the IBM research center in Rüschlikon,[4] Switzerland, and Buckminsterfullerenes (BUF), discovered in 1985 by Harold Kroto of the University of Sussex in the UK and Richard Smalley and Robert Curl of Rice University in Houston, Texas, USA. The development of STM was recognized by a Nobel Prize in Physics in 1986, and the discovery of BUF was recognized by a Nobel Prize in Chemistry in 1996.[5] By selecting these two research breakthroughs, we contribute to a long line of sociological studies on the Nobel Prize, its awardees, and research organizations recognized by Nobel prizes.[6]

Based on the selection of STM and BUF, we examined which universities seize upon such breakthroughs and how quickly they engage in follow-up research. Our analysis focuses on explaining the differences in the speed with which these breakthroughs were taken up and institutionalized within organizational units of the various universities. In this regard, we compare state universities in Germany with state universities in the USA. These two countries were the most important global centers of research in the late nineteenth century and all of the twentieth century.[7] However, the leading role of German universities in most scientific disciplines had been increasingly challenged by US universities since 1900, and in the 1930s, Germany was replaced by the USA as the new global center in scientific research. Today, in both countries, we find a lively scholarly and public policy discourse about academic leadership and excellence.[8]

For a meaningful comparison of the university systems in Germany and the USA, it has to be taken into account that the majority of German universities are funded by the Länder states. According to the classification of the German Federal Statistical Office (2010), 102 German universities have the right to award doctoral degrees, and 82 of these universities are state sponsored. In comparison, according to the Carnegie Classification (2010), 265 US universities have the right to award doctoral degrees, and 155 of them are state sponsored. Therefore, our comparison includes 82 German and 155 US state universities.

From a methodical point of view, the focus on state universities is important because it allows a direct comparison between the two countries. Taking into account the many private US universities funded by multibillion dollar endowments, such as Stanford, Caltech, Harvard, Yale, Princeton, Chicago, Columbia, and MIT, would distort the comparison. Private US universities constitute a particular institutional sector in a stratified educational system and would thus require a separate comparative analysis. However, such a comparison with the 20 private German universities would be quite difficult because the latter are only of minor importance in science and engineering. In sum, when we speak of universities in the following discussion, we are always referring to state universities in the two countries.

Our bibliometric findings demonstrate that scientists at US universities were several years ahead of their colleagues at German universities in seizing upon STM and BUF. Based on a set of hypotheses, this chapter demonstrates that universities with budgets that grew and that had a high number of professors among their scientific staff in the years following major scientific breakthroughs were among the early adopters and thus highly competitive in the new and emerging fields. In contrast, universities with stagnating budgets and a low share of professors among their scientific staff were mostly among those that engaged in follow-up research relatively late. These findings are elaborated using both longitudinal staff and funding data and retrospective interviews with key actors involved in follow-up research in various universities. We identify major differences in the university systems of Germany and the USA.

The chapter is structured as follows. The next section introduces the theoretical framework, highlighting two processes of gradual institutional change that are particularly important for renewal in science (Sect. 2).

Then, we introduce method and data (Sect. 3) as well as hypotheses and describe both the dependent variable and the explanatory variables (Sect. 4). These sections are followed by a detailed comparison of state universities in Germany and the USA (Sects. 5 and 6). Finally, we sum up our findings and draw conclusions.

6.2   Theoretical Framework

The emergence and expansion of new research areas is typically discussed with respect to disciplinary specialization and institutional differentiation. In this perspective, intellectual renewal takes place within established academic disciplines and often leads to new subdisciplines.[9] Yet this view accounts only for the result of both intellectual and institutional reconfigurations and neglects the often protracted and conflict-laden processes involved in spinning off new fields of research. The processes themselves as well as the mechanisms that propagate them and eventually make possible the successful implementation of new research areas have not been broadly studied, and both processes and mechanisms of renewal in science thus are relatively unknown territory.[10]

In recent years, sociologists and political scientists interested in explaining historical shifts in welfare state institutions have developed the approach of *historical institutionalism* that addresses institutional change from both a theoretical and an empirical perspective. In particular, Kathleen Thelen, James Mahoney, and Wolfgang Streeck have shown that institutional change in advanced economies often takes place gradually but nevertheless can result in fundamental changes to existing institutional structures. Among the gradual change processes identified by Thelen, Mahoney, and Streeck,[11] two processes, *layering* and *displacement*, are of particular importance here. *Layering* means that new research capacities are created while prior research is continued at the same or an even higher scale. In this way, new research areas are added to the existing fields. *Displacement* occurs when the creation of new research areas requires shrinking existing research fields. Like in a zero-sum game, support for new research fields is related to abandoning capacities in existing fields.

The historical institutionalism literature assumes *layering* to be the least conflict-laden process of gradual change.[12] This insight can be directly translated to renewal in science. Investments in capacities for a new research field mean no direct loss for the establishment in existing fields and thus provide a comfortable situation for innovators and early adopters.

In contrast, displacement is more conflict prone: The gains of the new field are the losses of existing fields; therefore, the scientific establishment will wield all of its influence to prevent or at least postpone changes in the status quo. Hence, in cases where the scientific establishment has strong veto power, renewal can be actively resisted.[13]

Building up research capacities in new fields requires scientific staff and financial resources, which are necessary but not sufficient conditions. We assume that the professor is the most important staff category for intellectual renewal at universities. He or she represents the smallest organizational unit that can make the decision to seize upon and invest in new scientific opportunity. There are two mechanisms for *displacement* of research areas at the level of professors. First, a professor may decide to change research areas. Because of their status, professors are entitled but also expected to make such decisions independently whereas other scientific staff and students typically require permission. The second mechanism is recruitment, which leads to renewal because newly recruited professors are specialized in new areas. As long as the absolute number of professors at a university remains constant, personnel fluctuation can lead to *displacement* of research areas. If the number of professors grows, then there is room for *layering* of additional research areas.

Regarding financial resources, we distinguish between the two broad categories of basic institutional funding and competitive grant funding because they are linked to intellectual renewal in different ways. In Germany, professorships are typically endowed with basic funding for scientific staff, laboratories, and equipment, which still made up a large share of their research budget during the 1980s, the time of the STM and BUF breakthroughs. Basic funding is flexible in the sense that it is not earmarked for specific project objectives. As long as basic funding grows, there is always some amount for investment in new topics and research opportunities. On the other hand, basic funds are tied to professorial chairs; thus, there is competition among chair holders for available basic funding. In this way, stagnating basic funding means that *displacement* is the only option for renewal whereas growth in basic funding indicates possibilities for *layering*.

The category of competitive grant funding includes public and private grants as well as other external research money that is invested in research projects. Grants are linked to intellectual renewal because they drive scientists to seek opportunities for rapidly demonstrable scientific achievement. Furthermore, grants are additional external resources that do not threaten existing research areas in which universities have invested their basic funding. Depending on the time frame under which they are allocated, grants allow for more or less stable *layering* of new research areas. But grant funding is also linked to *displacement* processes. In the USA, professorial positions are typically not endowed with staff and equipment. Professors who are unsuccessful in obtaining grants are in fact forced to abandon their research after a short time and to take up more teaching or administrative duties. As a consequence, research areas that are no longer approved by peer review or funding agencies are rapidly displaced.

6.3   Method and Data

This chapter combines quantitative and qualitative information to explain how staff structure and funding conditions influence the speed of reception of novel scientific ideas. Our focus is on findings of four case studies of universities that engaged in follow-up research of STM (two cases) and BUF (two cases). Each case was investigated in depth to find out how the influence of staff structure and funding resources played out in this particular instance of follow-up research. Summaries of case findings are organized according to the selected variables. Scientists who were interviewed are mentioned for each case study (see endnotes).

To draw generalizations from individual cases, we embedded each case in two longitudinal data sets. These data allow for systematic comparisons between cases and between the case and macro levels. The basis of the study is the construction of a strictly comparable set of state universities. The first data set consists of a bibliometric analysis of all state universities in Germany and the USA that engage in follow-up research for STM and BUF. Building on the available secondary literature on STM[14] and BUF[15] we used publication and citation data retrieved on the basis of "article flags" in Web of Science to investigate how rapid and how sustained the reception of these two breakthroughs was globally.[16]

The second macro data set consists of long-term personnel and funding data on the department, university, and state levels, which allows for a comparative analysis of institutional conditions for *layering* versus *displacement* of new research areas. These data were retrieved from the Bavarian Statistical Office, the University of California's Office of the President, the Integrated Postsecondary Education Data System, and further archival data from University of California, Santa Barbara (UCSB) and University of California, Los Angeles (UCLA). All funding data were inflation adjusted. To make scientific staff data for US universities comparable to scientific staff data for Bavarian universities, we used information on PhD graduates in US universities as a proxy for the number of scientific staff below the professoriate, the equivalent of what is called "wissenschaftliche Mitarbeiter" (scientific nonprofessorial staff) in German universities. Therefore, our values for the percentage of professors in US universities are lower and thus a stronger test compared to using raw data.

6.4   VARIABLES AND HYPOTHESES

The *dependent variable* in this analysis is the reception speed with which STM and BUF as research breakthroughs were taken up and expanded into research programs by scientists in state universities in Germany and the USA. Reception speed can be operationalized using the typology developed by Rogers (2003) for analyzing the diffusion of innovation. Rogers distinguishes *innovators*—that is, those who have achieved a scientific breakthrough—from *early adopters, early majority,* and *late majority*.[17] The early adopters are those scientists who promptly seize upon a breakthrough and adjust their own research to accommodate it; the early majority are those who get on board as the breakthrough begins to become accepted; and the late majority are those who join only in after the breakthrough has been widely adopted by peer scientists.

The analysis of STM and BUF follow-up research as documented below extends across and in part beyond 20 years. In the literature, it is common to conduct longitudinal analysis with either three- or five-year intervals.[18] We have chosen five-year periods. Accordingly, we define *early adopters* as those who started doing follow-up research within five years of the breakthrough; we define *early majority* as those who entered upon follow-up research in the second five-year period after the breakthrough; and *late majority* as those who started follow-up research more than ten years after the initial breakthrough, that is, in the third or fourth five-year period.

According to the theoretical framework outlined above, building up research capacities in new fields requires primarily scientific staff and appropriate funding. Therefore, we consider the following *explanatory variables*: relative frequency of professors among scientific staff, growth in the absolute number of professors, growth in absolute amount of basic funding, and percentage of grants in the funding structure (Table 6.1). These explanatory variables are outlined below.

Table 6.1   Hypotheses for explaining early adopters in STM and BUF

Hypothesis 1	Early adopters are found in universities with a high percentage of professors.
Hypothesis 2	Early adopters are found in universities with a growing number of professors.
Hypothesis 3	Early adopters are found in universities with growing basic funding.
Hypothesis 4	Early adopters are found in universities with a high percentage of grant funding.

The first explanatory variable is the percentage of professors among all scientific staff. It measures the extent to which universities host work units that are independent in making the decision to seize upon and invest in new scientific opportunities. Universities hosting many professors, relative to the entire scientific staff, are expected to have a short response time to research breakthroughs (hypothesis 1). This is for the following two reasons: First, hosting many professors raises the frequency by which new and emerging research opportunities are both detected and followed up by incumbent professors. Second, in any university, existing research areas are being replaced to some extent through staff fluctuation. Hosting many professors raises the frequency by which new professors are being hired, and new research topics and areas thus are imported. Therefore, the first explanatory variable is a measure of *displacement* of research areas.

In addition, the first explanatory variable is also an indicator for the average size of research groups. It carries information about working conditions and the leadership and management duties that are linked to the professorial position. According to previous research, small groups offer better environments for creative research because the group leader remains personally involved in research and because there is more frequent, more intensive, and less hierarchical communication between group leader and group members.[19] Doctoral students and postdocs in small group environments benefit from more intensive mentoring, which has been shown to be the best preparation for a successful academic career.[20] In contrast, in large groups, a professor is more involved in research management, which includes directing and supervising the implementation of a research program, acquisition and administration of grants, and more heavy coordinator and representative tasks in relation to scientific colleagues, university administration, and funding agencies. The cited advantages of small groups suggest they will on average show faster reception to new scientific ideas (*exploration*) whereas large groups enable more in-depth exploitation of already established scientific breakthroughs (*exploitation*).

The second explanatory variable is growth in the number of professors. It is interpreted as an indicator for processes of *layering* of new research areas. Recruitment of new professors is important for intellectual renewal because they are specialized in new areas. If the number of professors is growing, then recruitment frequency is above the replacement rate. In this situation, there will be less conflict and less resistance against the uptake of new research fields because there are more areas to add than to replace. Therefore, response time to novel scientific ideas is expected to be short when the number of professors is growing compared to universities where it is stagnating or declining over longer periods of time (hypothesis 2).

The third explanatory variable is growth of basic funding. Similar to growth in the number of professors, this variable measures processes of *layering*. Growth in basic funding means there are resources available for investment in new topics and research opportunities. Hence, research groups disposing of growing basic funding can react to new scientific developments swiftly and in a flexible manner. Scientists who work in the context of growing basic funding will—on average—show fast receptions to novel scientific ideas (hypothesis 3).

The fourth explanatory variable is the amount of public and private grants as percentage of basic university funding. Grants drive scientists to seek opportunities for rapidly demonstrable scientific achievement. Depending on the time frame under which they are allocated, grants allow for more or less stable *layering* of new research areas. It seems likely that universities with high portions of grants will have short response time to novel scientific ideas (hypothesis 4).

6.5   Empirical Results I: Reception Speed in German and US Universities

Our comparison includes all universities where scientists publish—on average—at least one publication per year citing the "article flags" of either STM or BUF. Therefore, we define as *early adopters* those universities who had at least five STM or five BUF publications in the years 1983–1987 and 1986–1990, respectively. *Early majority* are those universities that in the second five-year period had at least five STM or BUF publications in

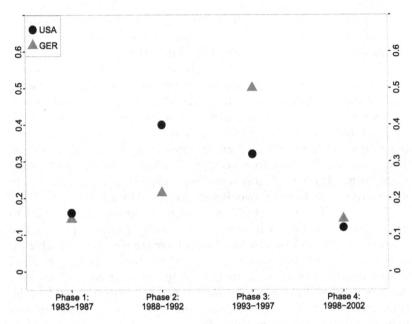

Fig. 6.1 Percentage of universities starting STM follow-up research
Source: WoS. Note: Phase 1 = Early adopters; Phase 2 = Early majority; Phases 3 and 4 = Late majority. N=14 German universities, N=25 US universities

1988–1992 and 1991–1995, respectively. *Late majority* comprised universities with an average of one publication per year and university more than ten years after the breakthrough, for STM in the years 1993–2002 and for BUF in the years 1996–2005.

The speed with which US universities compared to German universities entered follow-up research in STM and BUF is shown by their percentage in each of the five-year periods (Figs. 6.1 and 6.2). Regarding *early adopters*, there were four US universities and two German universities in STM, and six US universities but not a single German university in BUF. A second finding reinforces the first: Regarding *early majority*, there are mostly US universities, and the difference between US and German universities is more striking in BUF compared to STM. In contrast, German universities dominate in the category of *late majority*; the difference between the US and the German universities is again more striking in BUF compared to STM.

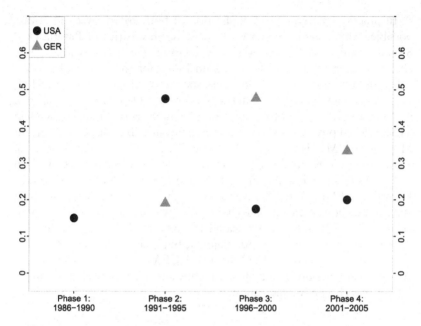

Fig. 6.2 Percentage of universities starting BUF follow-up research
Source: WoS. Note: Phase 1 = Early adopters; Phase 2 = Early majority; Phases 3 and 4 = Late majority. N=21 German universities, N=40 US universities

In sum, our bibliometric findings suggest that scientists in state universities in the USA were markedly ahead of their colleagues at German universities in seizing on both of these research breakthroughs. In the following section, we elaborate how these considerable differences in the dependent variable can be explained.

6.6   EMPIRICAL RESULTS II: CASE STUDIES OF GERMAN AND US UNIVERSITIES

Based on the bibliometric findings on the dependent variable, we established criteria for selecting university cases. For theoretical reasons, the first criterion was to choose universities that were either early adopters or early majority because the aim of the analysis is to determine which characteristics of our variables contribute to rapid follow-up research. A second criterion was the total number of STM or BUF publications that the universities published in the respective 20-year time frames.

In practice, the consistent application of both criteria was not always possible. The reason for this was that state universities in Bavaria (n = 8) and campuses of the University of California (UC; n = 10) had to be chosen because comparative longitudinal data for the independent variables could be retrieved only for these state universities. Regarding STM, Ludwigs-Maximilians-Universität München (LMU) was the first choice; in comparison to the other two Bavarian universities that engaged in STM follow-up research (Universität Regensburg, Technische Universität München), LMU is an early adopter and has a higher total number of STM publications. In the UC system, the choice was easy: UCSB is an early adopter and has the highest total number of STM publications. Regarding BUF, Friedrich-Alexander-Universität Erlangen-Nürnberg (FAU) was the first choice; like to the other two Bavarian universities that engaged in BUF follow-up research (Universität Bayreuth, Technische Universität München), it is late majority, but displays higher total numbers of BUF publications. In California, UCLA and UC Berkeley (UCB) are both early adopters, and almost identical in BUF publication output. UCLA was chosen for case study.

6.6.1   Explanatory Variables for UCSB, UCLA, LMU, and FAU

The first explanatory variable (V1) is the percentage of professors among all scientific staff. At UCSB, the percentage of professors decreased from 52% in the first period (1983–1988) to 42% in the last period (2003–2008). At UCSB's physics department, the percentage of professors decreased from 54% to 35% in the same periods. At UCLA, the percentage of professors slightly decreased from 44% in the first period (1986–1991) to 40% in the last period (2006–2010). At UCLA's chemistry and biochemistry department, the percentage of professors slightly decreased from 35% to 32% in the same periods.

At LMU, the percentage of professors decreased from 22% in the first period (1983–1988) to 12% in the last period (2003–2008). Figures for the physics department are almost identical with 21% in the first period and 13% in the last period. At FAU, the percentage of professors decreased from 19% in the first period (1986–1991) to 13% in the last period (2006–2010). Figures for the chemistry department are similar, with 19% in the first and 12% in the last period.

Compared to the US cases, the two German universities had a significantly lower percentage of professors among all scientific staff over the total observation period of 25 years, indicating a lower capacity for reception to novel scientific ideas. In addition, there is a general decrease in the percentage of professors in both systems, indicating a decreasing capacity for reception to novel scientific ideas.

The second explanatory variable (V2) is growth in the number of professors. At UCSB, this number rose by 44% (38% for full professors), from 531 in 1980 (311 full professors) to 767 in 2010 (505 full professors). At UCSB's physics department, their number rose by 48% (45% for full professors), from 25 (20 full professors) to 37 (29 full professors) in the same period. At UCLA, the total number of professors rose by 24% (35% for full professors) from 1267 in 1980 (741 full professors) to 1574 in 2010 (1001 full professors). At UCLA's chemistry and biochemistry department, however, their number rose by 9% only (24% for full professors), from 45 (29 full professors) to 49 (36 full professors) in the same period. Therefore, conditions at UCLA's chemistry and biochemistry department were less conducive than at UCLA in general.

At LMU, the number of professors was 906 in 1980 and decreased to 703 in 2010, a decline by 22% (including junior professors). The faculty of physics and astronomy had 38 professors in 1980. The figure rose to 42 in 1988, then stagnated (with minor fluctuations) until 2000, and then decreased to a minimum of 35 in 2006. Thus, there was slight growth in the first five years after the STM breakthrough. In connection with funding from the Deutsche Forschungsgemeinschaft (DFG) for a "cluster of excellence," professorial positions were in part reallocated by university leadership in the mid-2000s. The general declining trend was reversed for the physics department, and the number leaped to 51 professors in 2010. At FAU, the number of professors grew from 373 in 1980 to 524 in 2010, which is equivalent of a growth of 40%. In chemistry, there were 20 professors in 1980, and in physics there were 29. These figures remained roughly constant over 25 years, so that the uptake of BUF happened during a period of stagnation. Similar to the case of LMU, the launch of a DFG "cluster of excellence" led to noticeable growth at the end of the observation period, with 26 professorships in chemistry and 36 in physics in 2010.

The characteristics of the four universities on the two staff variables are quite typical of universities in the UC system and Bavaria. The UC system had a percentage of professors of 45% in the mid-1980s, declining to 40% in the second half of the 2000s. Bavaria had 22%, declining to 12% in the same period. In the UC system, the number of professors grew by 40% (71% for full professors), from 5155 in 1980 (2955 full professors) to 8552 in 2010 (5064 full professors). In Bavaria, the number of professors increased by 19%, from 2490 in 1980 to 2952 in 2010 (including junior professors). In absolute numbers, the UC system had 2.1 times as many professors as Bavaria in 1980 but 2.9 times as many in 2010. These differences in relative and absolute figures indicate an increasing divergence in the structure of scientific staff at UC campuses and Bavarian universities.

The third and fourth explanatory variables are growth in basic funding (V3) and percentage of state and private grant funding in total financial resources (V4). Basic state funding at UCSB's physics department shows long waves of growth, rising from $4.9 million in 1983 (first year of data set) to $7.0 million 1991, dropping to $5.1 million in 1994 and rising again to $8.0 million in 2004. Over a period of 28 years from 1983 to 2010, there was overall growth of 39% in basic state funding, and growth of 19% in tuition fees, both indicating overall good conditions for layering of new research areas. Furthermore, from 1983 to 2010, the amount of state and private grant funding oscillated between $2.8 and $4.5 million annually. As a result, between 1983 and 1987, the ratio between grant and basic funding at USCB's physics department fluctuated between 0.46 and 1.00, indicating very good conditions for layering of new research areas in the period following the STM breakthrough. In the entire observation period from 1983 to 2010, grant funding as a percentage of basic funding decreased from 46% (between 1983 and 1987) to 33% (between 2006 and 2010).

At the UCLA department of chemistry and biochemistry, basic state funding shows periods of decline and some growth in between. There was a decline from $14.0 million in 1986 (the first year of the data set) to $12.3 million in 1990, followed by a further decline to $11.4 million in 1995, then substantial growth to $13.8 million in 2000, followed by another decline to $12.8 million in 2010. However, the total decline of 8% in basic state funding between 1986 and 2010 was counterbalanced by a strongly increasing inflow from tuition fees, which more than doubled from $3.4 million in 1986 to $7.0 million in 2010. Therefore, total basic funding moderately increased by 14% between 1986 and 2010, indicating some opportunities for layering of new research areas. More layering possibilities existed because the amount of state and private grant funding rose by 43%, from $9.8 million in 1986 to $14.0 million in 2010. As a result, UCLA's chemistry and biochemistry department had a very high and increasing ratio of grant to basic funding, rising from 0.57 (between 1986 and 1990) to 0.9 (between 2006 and 2010).

At LMU, basic funding for the faculty of physics and astronomy declined in the period from 1982 to 1986 from around €19 million to €17 million. Later, basic funding rose to €21 million in 1987, and then declined again to €15 million in 1994, then rose again to €22 million in 2003. This means that during the 1980s and in the first half of the 1990s there were no additional basic funds available for the layering of new research areas at LMU. State and private grant funding at the faculties of physics and astronomy grew at first slowly between 1980 and 1995, and then more rapidly from €3.6 million in 1995 to €9.5 million in 2006, after which it surged to a maximum of €19.8 million in 2009. The ratio of grant to basic funding increased steadily from 0.17 (between 1986 and 1990) to 0.48 (between 2006 and 2010). At the end of the 2000s, DFG excellence funding caused statistical outliers. During the first decade of STM follow-up research at LMU, from 1983 to 1992, the ratio between grant and basic funding was still below 0.2, indicating limited resources for layering of new research areas.

At FAU, basic funding for the department of chemistry rose from €8 million in 1980 to €14.4 million in 1997 and then decreased to €11 million in 2009. The decline in basic funding since 1997 meant that no additional basic funds were available for the layering of new research areas at the time when BUF was taken up at FAU. State and private grant funding at the faculty of chemistry rose from €0.4 million in 1980 to €4.2 million in 2005. It fluctuated in the second half of the 2000s and reached a maximum of €4.9 million in 2009. The ratio of grant to basic funding increased slowly at first, from 0.03 (from 1986 to 1990) to 0.10 (from 1991 to 1995), and then sharply to 0.36 (from 2001 to 2005). In total, increasing grant funding compensated for the decline in basic funding from 1997 to 2004. Rising shares of grant funding showed overall good conditions for the layering of new research areas during the period when BUF was taken up at FAU.

The characteristics of the four universities on V3 and V4 are in many ways typical of the universities in the UC system and Bavaria. Basic funding for UC campuses grew from a total of $2.58 billion in 1979 to a total of $4.93 billion in 2010, which is equivalent of a growth of 91% (V3, including endowment). A decomposed analysis shows that state basic funding grew only slightly, by 12%, from $2.15 billion in 1980 to $2.41 billion in 2010, whereas tuition fees increased by a factor of 6.16, from $0.36 billion in

1979 to $2.65 billion in 2010. Even though state basic funding did not grow much, rising tuition fees led to pronounced long-term growth in basic funding, supporting continued layering of new research areas over a period of 30 years. Furthermore, state and private grant funding grew by a factor of 4.5, from $1.40 billion in 1979 to $6.32 billion in 2010. Therefore, the ratio of grant to basic funding (V4) increased from 0.54 in 1979 to 1.28 in 2010, indicating excellent conditions for the layering of new research areas.

Basic funding for state universities in Bavaria grew by 54%, from €1.06 billion in 1980 to €1.63 billion in 2010 (V3). This includes tuition fees, which were introduced in the second half of the 2000s, rising from €7.8 million in 2006 to €111.4 million in 2010. The growth period extends from 1980 to 1992; afterwards, there was a period of stagnation with fluctuations until 2007. Therefore, there were good conditions for layering of new research areas until the early 1990s, followed by a period of stagnation during the 1990s and 2000s. State and private grant funding expanded strongly by a factor of 7.2, from €80.3 million in 1980 to €577 million in 2010. The ratio of grant to basic funding (V4) increased from 0.08 in 1980 to 0.35 in 2010. Although the steep growth in grant funding indicates improving conditions for layering of new research areas, the percentage of grant funding was much lower compared to the UC system.

The analysis of staff structure and financial resources shows that, apart from minor deviations, the four selected cases are representative of macro developments in the respective state university systems. The quantitative description already hints at dramatic differences in the conditions for intellectual renewal in California and Bavaria. These differences are further investigated in each of the four case studies below.

6.6.2   UCSB (STM)[21]

The story of STM adoption at UCSB is the story of the Paul Hansma laboratory. Hansma is a physicist and early adopter who stepped into STM research in 1983 when Gerd Binnig for the first time presented atomic resolution images of a 7-by-7 silicon surface reconstruction (dependent variable). Before adopting STM, Hansma had worked on inelastic electron tunneling spectroscopy and already had been introduced to STM through personal contact with Binnig in the summer of 1981, a few months after the initial discovery. Hansma was also among the earliest adopters of the Atomic Force Microscope (AFM), as he shifted his research group from STM to AFM immediately after the invention by Binnig, Quate, and Gerber in 1986. During the 1990s, his group invented applications of AFM for a variety of disciplines, while in the 2000s, the focus shifted to development of biomedical AFM applications and devising improved diagnostics for skeleton bones.

The case of UCSB highlights the percentage of professors (V1) as a significant factor for the rapid uptake of research breakthroughs. The Hansma laboratory at UCSB represents an organizational structure geared to the individual investigator and his scientific collaborations. As a group leader, Hansma appreciates the advantages of small groups, and he cares to protect his own role as a researcher against encroachment by research management duties. As Hansma emphasized in an interview, he never wanted his group to become too big for himself to work in the laboratory or build prototypes with his own hands. Hansma became known in the "instrumental community" for recruiting a long series of postdocs who expanded STM and AFM applications into broad areas of physics, chemistry, materials science, geology, and molecular biology.[22] Over time, he collaborated with a large number of scientists from physics as well as other disciplines inside and outside UCSB. One of his most important partners in AFM research was Hermann Gaub, who stayed at UCSB as a postdoc in 1988 and became a professor at LMU in 1995, creating a substantive link between the two case studies.

Given that research groups are small, the main management duty of a professor consists in the acquisition of grants (V4). As described in the previous section, the physics department at UCSB had a high share of grant funding in the period between 1983 and 1987. The case study shows that Hansma used three approaches to secure flexible long-term funding for his group. First, he was able to obtain long-term grants, most of the time from the US National Science Foundation (NSF) Division of Materials Research, a grant that was extended four times over 30 years from 1973 to 2004, and later from the US National Institutes of Health from 2002 to 2014. A second parallel funding stream was provided by grants of shorter duration from varying sources.

Second, a strong reputation allowed Hansma to adopt the principle that he would accept only postdocs who brought their own funding with them. In that way, he reduced his own acquisition load while selecting postdocs who were capable of writing grant applications independently and whose projects could stand up to peer review. Third, a close collaboration with the start-up Digital Instruments Inc., founded by UCSB physics professor Virgil Elings in 1986, provided the Hansma laboratory with significant contributions in instrumentation and patent royalties, which he could use as flexible research money. Flexible as opposed to earmarked funding is important for reception speed to novel scientific ideas.

Strong growth in the number of professors at the physics department and at UCSB more generally (V2) underpins a recruiting policy geared at individual talent. We found no institutional commitment on the part of UCSB to build up or maintain excellence in STM/AFM research. Rather, UCSB aims to recruit the best and most talented individuals while it is understood that as professors, they may decide to change research areas perhaps several times over the course of their careers. Renewal is implemented as individual reaction to opportunity (V1).

Another interesting finding from the case of UCSB concerns the investment of additional basic funding (V3) and grant funding (V4) for shared resources that are accessible to all scientists either within the same department or across several departments. According to Hansma, the physics machine shop was most significant to the success of his group because there were excellent machinists who built instruments for researchers, and professors and students could also build things for themselves. In the 1980s, the physics department still partially covered the costs of the machine shop. Today, this machine shop is financed from individual research grants (V4) on a full cost basis. Still, the same infrastructure is provided for all scientists in the department of physics. Another example is the Materials Research Laboratory, which was established at UCSB in 1992 under the framework of the NSF's "Materials Research Science & Engineering Centers" (MRSEC) program. The MRSEC seeks to reinforce the base of individual investigator and small group research (V1) by supporting research approaches of a scope and complexity that would not be feasible under traditional funding of individual research projects. In this context, Hansma formed a long-lasting interdisciplinary collaboration with Galen Stucky, Daniel E. Morse, and later J. Herbert Waite. The MRSEC combines project grants (V4) for interdisciplinary teams of professors with the provision of facilities that are shared among members of different departments (V3). In this way, collaboration among faculty is facilitated.

In sum, the UCSB case demonstrates that along with the high percentage of professors among scientific staff positions, the professor and his small group are the key unit of decision making and thus of change in science (V1). In addition, *layering* new research areas requires long-term availability of individual investigator grants (V4) and sharing equipment and laboratory space via basic departmental or university funding (V3).

6.6.3   UCLA (BUF)[23]

The story of BUF adoption at UCLA involves the research groups of Robert Whetten, François Diederich, Richard Kaner, and Karoly Holczer. Their groups were among the early adopters of BUF research. The first phase of BUF follow-up research lasted from 1985 until 1990, when Krätschmer, Lamb, Fostiropoulos, and Huffmann introduced a new process for the synthesis of C60 molecules.[24] In 1990, when Whetten heard Krätschmer lecture on the C60 manufacturing processes at a conference in Germany, he immediately paid him a visit at the Max Planck Institute for Nuclear Physics in Heidelberg and, together with Diederich, started to produce C60 at UCLA. Whetten and Diederich were thus among the first scientists worldwide to enter the race for the chemical characterization of fullerenes (dependent variable). Together with Kaner and Holczer, they formed a team of complementary specialists and quickly attained a central position in the emerging field. In the period between May 1991 and September 1993, Whetten, Kaner, and Holczer co-authored 19 articles while Whetten and Diederich had another 20 co-publications.

Even though the percentage of professors in UCLA's chemistry and biochemistry department was lower than at UCLA in general, the case study illustrates the advantage of early scientific independence, which is linked to a high percentage of professors among scientific personnel (V1). Whetten was born in 1959 and thus barely over 30 years old when he stepped into C60 research. By the age of 26, he had already been an assistant professor. Diederich was born in 1952, and by age 33, he had completed his habilitation at Heidelberg before coming to UCLA in 1985. Despite the fact that Diederich was comparatively young when he completed his habilitation, he attained an independent research position seven years later than Whetten. As Kaner explained in an interview, the US system offers scientists the opportunity to succeed or fail at a very young age. Well below the age of 30, scientists may be given a laboratory with the equipment, students, and resources necessary to do whatever they are capable of doing. In contrast, their peers in Germany would typically work under supervision of a more established professor until their late thirties and early forties.[25]

The case of UCLA also illustrates how tenure track is linked to the acquisition of grant funding (V4). The tenure-track system works as an incentive structure that rewards rapid uptake of new scientific opportunity. When Diederich, Whetten, and Kaner stepped into BUF research in 1990, Diederich had shortly before been promoted to full professor, Whetten was an associate professor, and Kaner was an assistant professor. At the time of their appointment, they had been equipped with substantial starting capital from UCLA. As Kaner explained in an interview, Whetten advised him to expend his starting capital and more in order to earn scientific credit. Consequently, Whetten and Kaner both followed a deficit-spending strategy, consisting of rapid investment to come up with findings that would expedite the acquisition of new grant money. Judged by the criteria of the tenure track process, their strategy paid off. The scientific visibility and reputation that the group achieved in the initial BUF boom phase earned them rapid promotion to the status of full professor. Yet it was also risky because newly acquired research grants had to be used to settle previous debts, and the future revenue in external funding was never certain. Kaner was relieved from deficit spending in 1989 when he obtained a Hewlett Packard Fellowship worth $100,000 per year for a period of five years. Whetten, on the other hand, believed in the deficit-spending philosophy, and up until 1993, when he left UCLA, had accumulated massive debts on university accounts.

The case of UCLA also illustrates how the strong dependency of professors on grant funding (V4) may end a successful scientific collaboration. An apex of follow-up research at UCLA was the isolation of potassium-doped C60 compounds, demonstration of a single superconducting phase, and analysis of the crystal structure of K3C60. These findings were published in a race for priority with a group from Bell Labs. At the height of productivity, however, the collaboration disintegrated. In 1992, Diederich left UCLA for a professorial chair at Eidgenössische Technische Hochschule (ETH) Zürich; in 1993, Whetten accepted a professorship at the Georgia Institute of Technology in Atlanta. Holczer was appointed professor at UCLA in 1993 but felt compelled to change research fields after Whetten had left. Kaner stayed to continue on at UCLA with fullerene research. Compared to the 1991–1992 peak, the number of BUF follow-up publications dropped significantly.

In US universities, professorships are not endowed with staff positions, so except for the starting package professors may receive when accepting a professorial position, the entire laboratory, scientific group, students, and equipment must be sponsored through research grants. Diederich left UCLA to establish a much larger institute based on more extensive basic funding at the ETH Zürich, (V4). Twenty years later, his laboratory has issued a total of over 660 publications, awarded 106 doctoral degrees, and hosted 94 postdocs, attesting to differences in group size that are linked to the percentage of professors among scientific staff (V1). Whetten accepted the offer from the Georgia Institute of Technology, which allowed him to pay the debts he had accrued during his work at UCLA. Thus, it was the pressure to acquire grant money (V4) in a general climate of declining basic state funding that led to a premature disintegration of a highly productive collaboration in the case of UCLA's chemistry and biochemistry department.

In sum, the UCLA case demonstrates that the high percentage of professors among scientific staff (V1) made it possible for a team of four professors to build a coalition and, by means of some basic departmental and university funding (V3), but more importantly: by means of external grants (V4), successfully compete for a central position in the emerging research field. Although follow-up research at UCLA lasted from 1990 to 1993 only and thus shows that *layering* of new fields might be temporary, it was extremely productive during this period and represents an instance of rapid and successful response to novel scientific opportunity.

6.6.4   LMU (STM)[26]

STM follow-up research at LMU set in directly after the original breakthrough (dependent variable). This finding is not surprising given the fact that Gerd Binnig, one of the inventors of STM, came to LMU in 1987 as an honorary professor and for ten years led the IBM physics group there, an outpost of IBM Zürich. Other scientists involved in STM/AFM follow-up research include Wolfgang Heckl, Hermann Gaub, and Khaled Karrai.

Binnig set up his own laboratory at the institute of Theodor Hänsch, a physicist and pioneer of laser spectroscopy at LMU (Nobel laureate 2005). However, the IBM physics group seems to have exerted less influence than might be expected. This is displayed in a decreasing number of STM publications in the late 1980s and early 1990s. Binnig's title of honorary professor did not involve regular teaching duties or the right to supervise habilitations. As for his team, academic career options were either not readily available (V1, V2) or not attractive enough, so that most scientists moved on to other IBM projects and locations once the cooperation with LMU ended. Exceptions of team members who entered academia were Franz-Joseph Gießibl, who left the IBM physics group after his dissertation in 1991 and became a professor of experimental physics at Regensburg in 2006, and Wolfgang Heckl, who was a professor at LMU from 1993 to 2004.

The careers of Heckl and Gaub illustrate the scarcity of professorships (V1) and their decline in absolute numbers (V2) as a severe constraint on recruitment and thus on the uptake of new research areas at LMU in the late 1980s and 1990s. Heckl had been a doctoral student under Profs. Helmuth Möhwald and Erich Sackmann at the Institute of Biophysics at Technical University Munich (TUM) when Binnig recruited him. He joined the IBM physics group in 1989 as a postdoc. Because Binnig was not in a position to supervise his habilitation, Heckl became Hänsch's assistant in 1990 but continued to work with Binnig. The IBM laboratory was excellently equipped, and Heckl recalls a spirit of optimism and innovation there. Although his habilitation on the structure of DNA bases was awarded the Philip Morris Research Prize in 1993, at the age of 35, he could not be recruited to the physics faculty at LMU because between the late 1980s until the mid-2000s, the number of physics professors at LMU dropped from 42 (1988) to 35 (2005).

Therefore, he accepted an associate professor position for experimental physics at LMU's Institute of Crystallography in the faculty of geosciences in 1993. This move changed his working environment and conditions for the worse: He received little support among the full professors (chairholders) in geosciences, who perceived STM methods as unrelated to the core of their discipline. Because it is chairholders who are in the position to compete for and dispose of basic funding in German universities (V3), Heckl was left to finance his research group exclusively through external grants (V4). In 2004, Heckl was appointed director general at the German Museum in Munich. Even though his main responsibility there was science communication, he established an STM ultrahigh vacuum laboratory at the German Museum. In 2009, he was appointed full professor of science communication at TUM.

Gaub, like Heckl, had been a student of Prof. Sackmann and taken his doctorate in 1984 at TUM. He completed a postdoc at Stanford in 1984 and came to UCSB as a visiting scholar in 1988. There, he was introduced to AFM by Hansma, who handed him one of the first AFM prototypes.

The two scientists started a fruitful collaboration, co-publishing 12 papers on biophysical applications of AFM between 1990 and 1999. After Gaub had completed his habilitation and spent another year at Stanford, at the age of 38, he was appointed associate professor at TUM and in 1995 to full professor of applied physics at LMU. Gaub's recruitment to LMU was possible only because in 1995, the number of professors in physics almost reached the level of 1988 before it started to drop until 2005 again. Therefore, had there been more and a growing number of professor positions at LMU, Gaub, whom Hansma referred to as one of the most talented scientists he had ever collaborated with, could have possibly been recruited there much earlier.

Another finding concerns an institutional constraint on collaboration among faculty at LMU. Although the rise of the nanosciences since the early 1990s created a strong need for interdisciplinary collaboration among subspecialties of physics and other disciplines, professorial chairs at LMU showed little inclination for scientific exchange and collaboration because they competed individually for additional basic funding of their own chair-based research institutes that were operated by chairholders as self-contained hierarchical units (V3). In this situation, semiconductor physicist Jörg Peter Kotthaus together with a group of younger colleagues at LMU, including Heckl and Karrai, among others, created the Center of Nanosciences (CeNS) in 1998. CeNS brought together scientists who would open the doors of their laboratories to their colleagues as a precondition for CeNS membership, modeled after Kotthaus' experience at UCSB's department of physics. This organizational innovation reportedly unleashed a spirit of enthusiasm. CeNS was, in fact, one of the first of several nanoscience centers that have since been created in Germany and the USA.

In sum, the LMU case study shows that despite the presence of nobel laureate Gerd Binnig at the faculty of physics, the reception of novel scientific ideas was constrained by a low percentage of professors among scientific staff (V1) and by declining absolute numbers of professors both at the faculty of physics and at LMU (V2) during the late 1980s until the mid-2000s (*displacement*). Therefore, the opportunities to recruit outstanding scientists in the emerging field of STM/AFM follow-up research were severely inhibited.

6.6.5   FAU (BUF)[27]

FAU entered BUF follow-up research ten years after the original break-through (late majority). The case study begins in 1995 when Andreas Hirsch was appointed full professor at the Institute of Organic Chemistry (dependent variable). Hirsch formed a close collaboration with computer chemist Timothy Clark and physical chemist Dirk Guldi in the area of carbon allotropes. Since the beginning of the 2000s, the number of professors involved in this new research area increased through strategic activities both at the department and FAU level. Today, BUF follow-up research at FAU covers carbon nanotubes and graphene as well as fullerenes and involves collaborations among the departments of chemistry, physics, and material sciences.

The FAU case again highlights recruitment of professors as a key mechanism for intellectual renewal and suggests that a low percentage of professors among scientific staff (V1) causes late adoption of breakthroughs. When Hirsch was appointed in 1995, five years after the invention of C60 mass synthesis by Krätschmer et al.,[28] he was the only professor at FAU who had any experience in BUF-related research. Similar to the case of LMU, Hirsch reimported the topic from UCSB, where he had stayed from 1990 to 1991 as a postdoc with Fred Wudl, one of the first adopters of BUF research worldwide. Even though Hirsch was among the first adopters of fullerene chemistry in Germany, he first had to complete his habilitation in Tübingen before being recruited to an associate professorial position in Karlsruhe in 1995 and then to a full professor position at FAU in the same year. At FAU, Hirsch swiftly formed a collaboration with computer chemist Clark, who had been professor at FAU since 1976, and physical chemist Guldi, who despite having completed his doctoral thesis in 1991, one year after Hirsch, was appointed full professor at FAU as late as 2004.

When fullerene research started at FAU, it did so in a context of stag-nating numbers of professors at the chemistry department, as well as the university as a whole (V2). However, during the mid-2000s, there was a unique opportunity for intellectual renewal. Within a period of only a few years, 100 full professorial positions and 58 associate professorial positions were open for recruitment due to massive retirement. Facing this rare opportunity, FAU university leadership started to build strategic clusters in selected research fields. During this time, Hirsch and Clark had already built a collaboration that received departmental and university level support. Since 2000, university leadership defined carbon allotropes as part of FAU's profile in the strategic field of new materials research. This university strategy resulted in the appointment of a total of ten professors with research specialties related to carbon allotropes, five in the department of chemistry and five together in the departments of physics and material sciences; this is equivalent to *displacement* of existing by new research areas. In the context of organizational restructuring in 2007–2008, another three full professor positions were created in the "Interdisciplinary Centre for Molecular Materials"; this is equivalent to *layering* of new research areas on top of existing ones.

The concentration of basic funding (V3) into carbon allotropes was dependent on the successful acquisition of grant funding, especially from DFG. The percentage of grant funding (V4) in the department of chemistry had increased slowly from 3% (from 1986 to 1990) to 10% (from 1991 to 1995). During the first decade of BUF follow-up research, it climbed to 20% (from 1996 to 2000). Hirsch and Clark had received individual investigator grants from DFG for BUF follow-up research since 1996. From 2001 to 2012, they both led research groups within the DFG collaborative research center "Redoxactive Metal Complexes" (SFB 583). Then Clark was among the coordinators for FAU's acquisition of a DFG "cluster of excellence" in the field of advanced materials, which involves professors from several disciplines. This cluster yielded €41 million from DFG and additional €41 million together from the state of Bavaria, the German federal government, and industry for a period of five years from 2007 to 2012. Between 2012 and 2017, DFG granted another €34 million. Hirsch is director of the DFG collaborative research center "Carbon Allotropes" (SFB 953) for the period 2012–2017, coordinating 15 research groups in the departments of chemistry, physics, and engineering. Therefore, the percentage of grant funding (V4) in the department of chemistry increased to 36% (from 2001 to 2005).

As argued in the case of LMU, the professorial chair system operating with a small percentage of full professorships (V1) who then compete for additional basic funds (V3) tends to create self-contained units that impede collaboration. At FAU, this problem was addressed in an organizational reform in 2007–2008: Departments were created as administrative units below the level of faculties, replacing the former disciplinary institutes. The main objective of this reform was to make university administration more efficient and to improve administrative services. The department structure has been cited in interviews as a facilitating condition for collaboration among professorial chairs. However, apart from the sharing of administrative resources, the hierarchical professorial chair system remained intact. At the end of the 2000s, funding from the DFG excellence program allowed for some growth in the number of professors (V2) in the departments of chemistry, physics, and material sciences. At the same time, however, the numbers of scientific staff rose from an already high level, resulting in still lower percentages of professors and increased average group size (V1). Therefore, it is expected that the DFG excellence program has not sped up today's reception time for more recent research breakthroughs compared to the 1980s and 1990s.

In sum, the FAU case study shows how the reception of novel scientific ideas was constrained by a low percentage of professors among scientific staff (V1) and by stagnating absolute numbers of professors (V2) during the 1990s until the mid-2000s. Intellectual renewal happened at FAU with considerable delay only when, because of a retirement wave, a considerable number of professorial positions were open for recruitment (*displacement*), and when the university leadership took this opportunity to build strategic research areas and at the same time invested additional resources (V3) in these new areas, including professorial positions (*layering*). It also illustrates how large-scale grant funding (V4) ignited the systematic exploitation of carbon allotropes as an already recognized and established research field.

6.7   Conclusions

This chapter examines the capabilities of universities to rapidly build up and expand research capacities in new and emerging scientific fields following major scientific breakthroughs. Based on STM and BUF, two research breakthroughs in physics and chemistry from the early/mid-1980s, we investigated how quickly scientists in German and US state universities built up follow-up research in response to these breakthroughs. Most importantly, we explored to what extent the institutional framework in which universities are embedded supported such expansion and renewal. For this purpose, we distinguished between *layering* and *displacement* as gradual processes of renewal in science. Using longitudinal staff and funding data as well as case study evidence, we have provided original insights into mechanisms shaping these two renewal processes.

Our bibliometric findings (dependent variable) demonstrate that scientists in US universities were several years ahead of their colleagues at

German universities in seizing on STM and BUF. US scientists were more often early adopters and early majority than German scientists while the latter were mostly late majority. Our institutional findings (explanatory variables) suggest that in the years following STM and BUF, the UC system provided better institutional conditions for scientific renewal than universities in Bavaria. Universities in the UC system had many opportunities for taking up new and emerging fields, mostly via *layering* of new resources, including additional professorial positions, and via *displacement* of old by new research specializations that came with continuous replacement of professorial positions in universities with a high share of such positions among all scientific staff. In contrast, Bavarian universities operated under less supportive conditions: stagnating basic funding primarily invested in hierarchical, self-contained professorial chairs in combination with a relatively low level of external grant funding and scarcity of professorial positions caused delayed responses to novel scientific developments. Below are our results:

First, a high percentage of professors among scientific staff (V1) is conducive to intellectual renewal via *displacement* of established fields by new research fields, as stated in the first hypothesis. Two mechanisms are involved: A high percentage of professors raises the frequency by which new research opportunities are both detected and followed up by those who are expected to conduct independent research; in addition, a high percentage of professors raises the frequency by which new peers are hired, and new research topics and areas thus are imported in exchange for existing ones.

As the four cases have shown, the percentage of professors provides valuable information about the chance structure for academic careers. A low percentage of professors, as in Germany, indicates that many more young scientists work in the academic system than can be possibly absorbed into professorial ranks. As a consequence, there is a bottleneck at the transition to professorial status, leading to prolonged periods of dependency and job insecurity in academic biographies. In the US system, the transition to assistant professor, and thus scientific independence, takes place earlier in the biography, thus providing favorable conditions for seizing upon new and promising scientific opportunities.

Second, the chapter demonstrates that an increasing number of professors (V2), growth in basic funding (V3), and a high ratio of grant to basic funding (V4) are key factors positively associated with renewal via *layering* of new research areas on top of existing commitments in established research fields and disciplines, as stated by the second, third, and fourth hypotheses. In fact, a declining or stagnating number of professors (V2) severely constrains the capability of universities and their departments to build up swiftly new and emerging research fields by recruiting outstanding scientists, as demonstrated in the cases of LMU and FAU. Furthermore, as the case of UCSB shows, if growth of basic funding (V3) is channeled into facilities and laboratories that are widely shared by professors both inside and across departments, opportunities for particularly effective collaborations in new and emerging fields are created. Yet, as the case of UCLA illustrates, in a context of declining basic state funding, too strong dependency of professors on grant money and too high competitive pressure for external research resources (V4) may inadvertently end successful scientific collaborations before all fruits are harvested.

Third, our findings point to considerable and increasing differences in the university systems of California and Bavaria with major implications for renewal in science. Although the percentage of professors (V1) has decreased in both states since the 1980s, this decrease has happened in very different ranges: from 45% to 40% in California, and from 22% to 12% in Bavaria. Therefore, given our empirical findings on V1, the conditions for renewal in science in Bavarian universities are worse today than they were in the 1980s, in contrast to California.

Furthermore, basic funding (V3) for UC campuses grew from a total of $2.58 billion in 1980 to $4.93 billion in 2010 (91% growth) with tuition fees and grant funding providing the lion's share in growth. In contrast, basic funding for state universities in Bavaria grew from €1.06 billion in 1980 to €1.63 billion in 2010 (54% growth), including tuition fees (since 2007). Yet, following a more general political trend against tuition fees in all German Länder states, the Bavarian parliament abolished tuition fees in 2013. Tuition fees will not be charged in the future, thus reducing the level of basic funding in Bavarian universities. Therefore, based on our empirical findings on V3, the financial conditions for renewal in science in Bavarian universities are worse than in California.

In addition, there is also a major gap regarding the share of grant funding (V4) between the two states. In UC campuses, state and private grant funding grew from $1.40 billion in 1980 to $6.32 billion in 2010 (growth factor of 4.5), which is equivalent to an increase from 0.54 to 1.28 of grant relative to basic funding. In Bavarian universities, state and private grant funding expanded from €80.3 million in 1980 to €577 million in 2010 (a factor of 7.2), which is equivalent to an increase from 0.08 to

0.35. While the steep growth in grant funding indicates improving conditions for layering of new research areas in Bavaria, the ratio of grant to basic funding is still much lower compared to the UC system. In fact, the growth of state and private grant funding in Bavaria seems decoupled from the growth in the number of professors (V2): In Bavaria, the number of professors has grown by 19% while UC campuses have a growth of 40%. In comparison with the growth in state and private grant funding (Bavaria: 7.2, UC system: 4.5), much of the grant funding in Bavaria is channeled into scientific staff positions below the professorial level, which is typically not entitled to conduct scientific research independently—a key condition for renewal in science, as this chapter has shown.

In methodological terms, the chapter has demonstrated that interpreting qualitative results from the four case studies requires triangulation with longitudinal quantitative data on staff structure and funding streams. Without these quantitative data, it would be difficult to generalize results. In fact, the four cases represent the two university systems so well that findings at both the department and the university levels oftentimes match with variables for the two systems as a whole. In this way, the chapter strives to link the historical narrative of particular cases with more general institutional developments in the systems in which these cases are embedded.

NOTES

1. Jerry Gaston, *Originality and Competition in Science* (Chicago: Chicago University Press, 1973); Michael Polanyi, *Knowing And Being*. With an Introduction by Marjorie Grene (Chicago: Chicago University Press, 1969); Richard Whitley, *The Intellectual and Social Organization of the Sciences: Second Edition* (Oxford: Oxford University Press, 2000).
2. James G. March, "Exploration and Exploitation in Organizational Learning," *Organization Science* 2 (1991): 71–87.
3. Jerald Hage, and Jonathon Mote, "Transformational organizations and institutional change: the case of the Institut Pasteur and French science," *Socio-Economic Review* 6 (2008): 313–336; Thomas Heinze, "Creative Accomplishments in Science: Definition, Theoretical Considerations, Examples from Science History, and Bibliometric Findings," *Scientometrics* 95 (2013): 927–940; Thomas Heinze et al., "Organizational and Institutional Influences on Creativity in Scientific Research," *Research Policy* 38 (2009): 610–623; J. Rogers Hollingsworth, "A Path-Dependent Perspective on Institutional and Organizational Factors Shaping Major

Scientific Discoveries," in *Innovation, Science, and Institutional Change*, ed. Jerald Hage and Marius Meeus (Oxford: Oxford University Press, 2006), 423–442; Grit Laudel and Jochen Gläser, "Beyond breakthrough research: Epistemic properties of research and their consequences for research funding," *Research Policy* 43 (2014): 1204–1216; Robert K. Merton and Elinor G. Barber, *The Travels and Adventures of Serendipity: A Study in Sociological Semantics and the Sociology of Science* (Princeton, N.J.: Princeton University Press, 2004); Jan Youtie et al., "Career-based Influences on Scientific Recognition in the United States and Europe: Longitudinal Evidence from Curriculum Vitae Data," *Research Policy* 42 (2013): 1341–1355.

4. Gerd Binnig and Heinrich Rohrer, "Scanning Tunneling Microscopy," *Surface Science* 126 (1982): 336–344.

5. Harold Kroto et al., "C-60: Buckminsterfullerene," *Nature*, 318 (1985): 162–163.

6. Max Haller, Birgit Wohinz and Margot Wohinz, *Österreichs Nobelpreisträger und Wissenschaftler im historischen und internationalen Vergleich* (Wien: Passagen Verlag, 2002); Thomas Heinze et al., "New Patterns of Scientific Growth? How Research Expanded after the Invention of Scanning Tunneling Microscopy and the Discovery of Buckminsterfullerenes," *Journal of the American Society for Information Science and Technology* 64 (2013): 829–843; J. Rogers Hollingsworth, *Research Organizations and Major Discoveries in Twentieth-century Science: a Case Study of Excellence in Biomedical Research* (Berlin: Wissenschaftszentrum Berlin für Sozialforschung, 2002); J. Rogers Hollingsworth, "Institutionalizing Excellence in Biomedical Research: The Case of The Rockefeller University," in *Creating a Tradition*, ed. D.H. Stapleton (New York: Rockefeller University Press, 2004), 17–63; Harriet Zuckerman, *Scientific Elite: Nobel Laureates in the United States* (New York: Free Press, 1977); Harriet Zuckerman, *Die Werdegänge von Nobelpreisträgern. In Generationsdynamik in der Forschung* (Frankfurt am Main: Campus, 1993).

7. Joseph Ben-David, *The Scientist's Role in Society* (Chicago: University of Chicago Press, 1971); Tim Lenoir, *Instituting Science: The Cultural Production of Scientific Disciplines* (Stanford, California: Stanford University Press, 1997).

8. Mitchell G. Ash, *Mythos Humboldt: Vergangenheit und Zukunft der deutschen Universitäten* (Wien: Böhlau, 1997); Burton R. Clark, *Creating Entrepreneurial Universities: Organizational Pathways of Transformation* (New York: Pergamon Press, 1998); Jonathan R. Cole, *The Great American University* (New York: Public Affairs, 2009); David John Frank and Jay Gabler, *Reconstructing the University: Worldwide Shifts in Academia in the*

20th Century (Stanford: Stanford University Press, 2006); Reinhard Kreckel, *Zwischen Promotion und Professur: Das wissenschaftliche Personal in Deutschland im Vergleich mit Frankreich, Großbritannien, USA, Schweden, den Niederlanden, Osterreich und der Schweiz* (Leipzig: Akademische Verlagsanstalt, 2008); Georg Krücken et al., *Towards a Multiversity? Universities between Global Trends and National Traditions* (Bielefeld: Transcript, 2007); Richard Münch, *Die akademische Elite. Zur sozialen Konstruktion wissenschaftlicher Exzellenz* (Frankfurt am Main: Suhrkamp, 2007); Jack H. Schuster and Martin J. Finkelstein, *The American Faculty: The Restructuring of Academic Work and Careers* (Baltimore: Johns Hopkins University, 2006); Ulrich Teichler, *Higher Education Systems* (Rotterdam/Taipeh: Sense Publishers, 2007).

9. Rudolf Stichweh, *Wissenschaft, Universität, Professionen: Soziologische Analysen* (Frankfurt am Main: Suhrkamp, 1993); Peter Weingart, *Wissenschaftssoziologie* (Bielefeld: Transkript, 2003); Whitley, *The Intellectual and Social Organization.*

10. As an example of an early attempt, see Michael Joseph Mulkay, "Conformity and Innovation in Science," *The Sociological Review*, 18 (1972): 5–23; Michael Joseph Mulkay, "Three Models of Scientific Development," *The Sociological Review*, 23 (1975): 509–526.

11. James Mahoney and Kathleen Thelen, "A Theory of Gradual Institutional Change," in *Explaining Institutional Change: Ambiguity, Agency, and Power*, ed. James Mahoney and Kathleen Thelen (Cambridge: Cambridge University Press, 2010), 1–37; Wolfgang Streeck and Kathleen Thelen, "Introduction," in *Beyond Continuity. Institutional Change in Advanced Political Economies*, ed. Wolfgang Streeck and Kathleen Thelen (Oxford: Oxford University Press, 2005), 1–39; Kathleen Thelen, "How Institutions Evolve. Insights from Comparative Historical Analysis," in *Comparative Historical Analysis in the Social Sciences*, ed. James Mahoney and Dietrich Rueschemeyer (New York: Cambridge University Press, 2003), 208–240.

12. Streeck, Thelen, "Introduction"; Thelen, "How Institutions Evolve".

13. Duncon Wilson, *Reconfiguring Biological Sciences in the Late Twentieth Century: A Study of the University of Manchester* (Manchester: University of Manchester, 2008).

14. Chunli Bai, *Scanning Tunneling Microscopy and Its Applications* (New York: Springer, 2000); C. Julian Chen, *Introduction to Scanning Tunneling Microscopy* (New York: Oxford University Press, 1993); Hyungsub Choi and Cyrus C.M. Mody, "The Long History of Molecular Electronics: Microelectronics Origins of Nanotechnology," *Social Studies of Science*, 39 (2009): 11–50; Arne Hessenbruch, "Nanotechnology and the Negotiation of Novelty," in *Discovering the Nanoscale*, ed. Davis Baird, Alfred Nordmann and Joachim Schummer (Amsterdam: IOS Press, 2004), 135–

144; Cyrus C.M. Mody, *Instrumental Community: Probe Microscopy and the Path to Nanotechnology* (Cambridge, MA: MIT Press, 2011).

15. Hugh Aldersey-Williams, *The Most Beautiful Molecule: An Adventure in Chemistry* (London: Aurum Press, 1995); Jim Baggot, *Perfect Symmetry: The Accidental Discovery of Buckminsterfullerene* (Oxford: Oxford University Press, 1994); Rudy M. Baum, "Ideas on Soot Formation Spark Controversy," *Chemical and Engineering News* 68 (1990): 30–32.

16. For details, see Heinze et al., "New Patterns of Scientific Growth".

17. Everett M. Rogers, *Diffusion of Innovations* (New York: Free Press, 2003).

18. Lee Fleming, Santiago Mingo and David Chen, "Collaborative Brokerage, Generative Creativity, and Creative Success," *Administrative Science Quarterly* 52 (2007): 443–475; Walter W. Powell et al., "Network Dynamics and Field Evolution: The Growth of Interorganizational Collaboration in the Life Sciences," *American Journal of Sociology* 110 (2005): 1132–1205.

19. Heinze et al., "Organizational and Institutional Influences"; Hollingsworth, "Institutionalizing Excellence"; Hollingsworth, "A Path-Dependent Perspective"; Youtie et al., "Career-based Influences".

20. J. Scott Long and Robert McGinnis, "Organizational Context and Scientific Productivity," *American Sociological Review* 46 (1981): 422–442; Youtie et al., "Career-based Influences"; Harriet Zuckerman, *Die Werdegänge von Nobelpreisträgern*.

21. Interviewees in alphabetical order: Guenther Ahlers, 20.3.2012; Tom Albrecht, 30.3.2012; Steve Buratto, 19.3.2012; Paul Hansma, 29.3.2012; Robert Madix, 3.4.2012; Fred Wudl, 22.3.2013; Joseph Zasadzinski, 22.3.2012.

22. Mody, *Instrumental Community*.

23. Interviewees in alphabetical order: Karoly Holczer, 29.3.2012; Richard Kaner, 20.3.2012; Robert Madix, 3.4.2012; Yves Rubin, 19.3.2012; Fred Wudl, 22.3.2012.

24. W. Krätschmer et al., "Solid C60: A New Form of Carbon," *Nature* 347 (1990): 345.

25. Jürgen Enders, *Die wissenschaftlichen Mitarbeiter: Ausbildung, Beschäftigung und Karriere der Nachwuchswissenschaftler und Mittelbauangehörigen an den Universitäten* (Frankfurt, New York: Campus Verlag,1996); Kreckel, *Zwischen Promotion und Professur.*

26. Interviewees in alphabetical order: Paul Hansma, 29.3.2012; Wolfgang Heckl, 28.2.2012; Khaled Karrei, 25.6.2012; Jörg Kotthaus, 15.10.2012; Stefan Thalhammer, 25.6.2012.

27. Interviewees in alphabetical order: Timothy Clark, 28.7.2011; Dirk M. Guldi, 26.9.2011; Frank Hauke, 22.9.2011; Andreas Hirsch, 5.6.2012, 19.4.2012, Norbert Jux, 13.7.2011.

28. W. Krätschmer et al., "Solid C60."

Acknowledgments This chapter is based on research that was supported by the German Federal Ministry for Research and Education (BMBF) grant 01UZ1001. We thank Richard Heidler and Heiko Heiberger for assistance in conducting interviews and retrieving statistical data at the Bavarian Statistical Office. We are very grateful to David Pithan for processing bibliometric data, Joel Fuchs for retrieving valuable additional personnel and funding data at archives at UCLA and UCSB, and at the German Federal Statistical Office, and for cleaning and processing all personnel and funding data. We are also very grateful for comments and suggestions from the participants of the Research Colloquium at the Institute of Sociology, Technical University Berlin (January 14, 2015), the 20th International Conference on Science and Technology Indicators in Lugano (September 2–4, 2015), and the Atlanta Conference on Science and Innovation Policy (September 17–19, 2015) where earlier versions of this chapter were presented.

References

Aldersey-Williams, Hugh. 1995. *The most beautiful molecule: An adventure in chemistry.* London: Aurum Press.

Ash, Mitchell G. 1997. *Mythos Humboldt: Vergangenheit und Zukunft der deutschen Universitäten.* Wien: Böhlau.

Baggot, Jim. 1994. *Perfect symmetry: The accidental discovery of buckminsterfullerene.* Oxford: Oxford University Press.

Bai, Chunli. 2000. *Scanning tunneling microscopy and its applications.* New York: Springer.

Baum, Rudy M. 1990. Ideas on soot formation spark controversy. *Chemical and Engineering News* 68: 30–32.

Ben-David, Joseph. 1971. *The scientist's role in society.* Chicago: University of Chicago Press.

Binnig, Gerd, and Heinrich Rohrer. 1982. Scanning tunneling microscopy. *Surface Science* 126: 336–344.

Chen, C. Julian. 1993. *Introduction to scanning tunneling microscopy.* New York: Oxford University Press.

Choi, Hyungsub, and Cyrus C.M. Mody. 2009. The long history of molecular electronics: Microelectronics origins of nanotechnology. *Social Studies of Science* 39: 11–50.

Clark, Burton R. 1998. *Creating entrepreneurial universities: Organizational pathways of transformation.* New York: Pergamon Press.

Cole, Jonathan R. 2009. *The Great American University.* New York: Public Affairs.

Enders, Jürgen. 1996. *Die wissenschaftlichen Mitarbeiter: Ausbildung, Beschäftigung und Karriere der Nachwuchswissenschaftler und Mittelbauangehörigen an den Universitäten.* Frankfurt/New York: Campus Verlag.

Fleming, Lee, Santiago Mingo, and David Chen. 2007. Collaborative brokerage, generative creativity, and creative success. *Administrative Science Quarterly* 52: 443–475.

Frank, David John, and Jay Gabler. 2006. *Reconstructing the university: Worldwide shifts in academia in the 20th century*. Stanford: Stanford University Press.

Gaston, Jerry. 1973. *Originality and competition in science*. Chicago: Chicago University Press.

Hage, Jerald, and Jonathon Mote. 2008. Transformational organizations and institutional change: The case of the Institut Pasteur and French science. *Socio-Economic Review* 6: 313–336.

Haller, Max, Birgit Wohinz, and Margot Wohinz. 2002. *Österreichs Nobelpreisträger und Wissenschaftler im historischen und internationalen Vergleich*. Wien: Passagen Verlag.

Heinze, Thomas. 2013. Creative accomplishments in science: Definition, theoretical considerations, examples from science history, and bibliometric findings. *Scientometrics* 95: 927–940.

Heinze, Thomas, and Georg Krücken. 2012. *Institutionelle Erneuerungsfähigkeit der Forschung*. Wiesbaden: Springer VS.

Heinze, Thomas, Philip Shapira, Juan D. Rogers, and Jacqueline M. Senker. 2009. Organizational and institutional influences on creativity in scientific research. *Research Policy* 38: 610–623.

Heinze, Thomas, Richard Heidler, Heiko Heiberger, and Jan Riebling. 2013. New patterns of scientific growth? How research expanded after the invention of scanning tunneling microscopy and the discovery of Buckminsterfullerenes. *Journal of the American Society for Information Science and Technology* 64: 829–843.

Hessenbruch, Arne. 2004. Nanotechnology and the negotiation of novelty. In *Discovering the nanoscale*, ed. Davis Baird, Alfred Nordmann, and Joachim Schummer, 135–144. Amsterdam: IOS Press.

Hollingsworth, J. Rogers. 2004. Institutionalizing excellence in biomedical research: The case of The Rockefeller University. In *Creating a tradition*, ed. D.H. Stapleton, 17–63. New York: Rockefeller University Press.

Hollingsworth, J. Rogers. 2006. A path-dependent perspective on institutional and organizational factors shaping major scientific discoveries. In *Innovation, science, and institutional change*, ed. Jerald Hage and Marius Meeus, 423–442. Oxford: Oxford University Press.

Krätschmer, W., L.D. Lamb, K. Fostiropoulos, and D.R. Huffman. 1990. Solid C60: A new form of carbon. *Nature* 347: 345.

Kreckel, Reinhard. 2008. *Zwischen Promotion und Professur: Das wissenschaftliche Personal in Deutschland im Vergleich mit Frankreich, Großbritannien, USA, Schweden, den Niederlanden, Österreich und der Schweiz*. Leipzig: Akademische Verlagsanstalt.

Kroto, H.W., J.R. Heath, S.C. O'Brien, R.F. Curl, and R.E. Smalley. 1985. C-60: Buckminsterfullerene. *Nature* 318: 162–163.

Krücken, Georg, Anna Kosmützky, and Mark Torka. 2007. *Towards a multiversity? Universities between global trends and national traditions.* Bielefeld: Transcript.

Laudel, Grit, and Jochen Gläser. 2014. Beyond breakthrough research: Epistemic properties of research and their consequences for research funding. *Research Policy* 43: 1204–1216.

Lenoir, Tim. 1997. *Instituting science: The cultural production of scientific disciplines.* Stanford: Stanford University Press.

Long, J. Scott, and Robert McGinnis. 1981. Organizational context and scientific productivity. *American Sociological Review* 46: 422–442.

Mahoney, James, and Kathleen Thelen. 2010. A theory of gradual institutional change. In *Explaining institutional change: Ambiguity, agency, and power*, ed. James Mahoney and Kathleen Thelen, 1–37. Cambridge: Cambridge University Press.

March, James G. 1991. Exploration and exploitation in organizational learning. *Organization Science* 2: 71–87.

Merton, Robert K., and Elinor G. Barber. 2004. *The travels and adventures of serendipity: A study in sociological semantics and the sociology of science.* Princeton: Princeton University Press.

Mody, Cyrus C.M. 2011. *Instrumental community: Probe microscopy and the path to nanotechnology.* Cambridge, MA: MIT Press.

Mulkay, Michael J. 1972. Conformity and innovation in science. *The Sociological Review* 18: 5–23.

Mulkay, Michael J. 1975. Three models of scientific development. *The Sociological Review* 23: 509–526.

Münch, Richard. 2007. *Die akademische Elite. Zur sozialen Konstruktion wissenschaftlicher Exzellenz.* Frankfurt am Main: Suhrkamp.

Polanyi, Michael. 1969. *Knowing and being.* With an Introduction by Marjorie Grene. Chicago: Chicago University Press.

Powell, Walter W., Douglas R. White, Kenneth W. Koput, and Jason Owen-Smith. 2005. Network dynamics and field evolution: The growth of interorganizational collaboration in the life sciences. *American Journal of Sociology* 110: 1132–1205.

Rogers, Everett M. 2003. *Diffusion of innovations.* New York: Free Press.

Schuster, Jack H., and Martin J. Finkelstein. 2006. *The American faculty: The restructuring of academic work and careers.* Baltimore: Johns Hopkins University.

Servos, John W. 1990. *Physical chemistry from Ostwald to Pauling: The making of a science in America.* Princeton: Princeton University Press.

Stichweh, Rudolf. 1993. *Wissenschaft, Universität, Professionen: Soziologische Analysen*. Frankfurt am Main: Suhrkamp.

Streeck, Wolfgang, and Kathleen Thelen. 2005. Introduction. In *Beyond continuity. Institutional change in advanced political economies*, ed. Wolfgang Streeck and Kathleen Thelen, 1–39. Oxford: Oxford University Press.

Teichler, Ulrich. 2007. *Higher education systems*. Rotterdam/Taipeh: Sense Publishers.

Thelen, Kathleen. 2003. How institutions evolve. Insights from comparative historical analysis. In *Comparative historical analysis in the social sciences*, ed. James Mahoney and Dietrich Rueschemeyer, 208–240. New York: Cambridge University Press.

Weingart, Peter. 2003. *Wissenschaftssoziologie*. Bielefeld: Transkript.

Whitley, Richard. 2000. *The intellectual and social organization of the science*, 2nd ed. Oxford: Oxford University Press.

Wilson, Duncan. 2008. *Reconfiguring biological sciences in the late twentieth century: A study of the University of Manchester*. Manchester: University of Manchester.

Youtie, Jan, Juan Rogers, Thomas Heinze, Philip Shapira, and Li Tang. 2013. Career-based influences on scientific recognition in the United States and Europe: Longitudinal evidence from curriculum vitae data. *Research Policy* 42: 1341–1355.

Zuckerman, Harriet. 1977. *Scientific elite: Nobel laureates in the United States*. New York: Free Press.

Zuckerman, Harriet. 1993. Die Werdegänge von Nobelpreisträgernn. In *Generations- dynamik in der Forschung*. ed. Karl Ulrich Mayer, 59–79. Frankfurt am Main: Campus.

CHAPTER 7

Organizing Space: Dutch Space Science Between Astronomy, Industry, and the Government

David Baneke

7.1   INTRODUCTION

Whenever a new technological or scientific field emerged after the Second World War, Dutch scientists, government officials, and industrial companies feared being left behind. Especially in strategically important fields such as nuclear physics, radio astronomy (radar), and computing, scientists, industrial companies, and the government cooperated to initiate research efforts. These cooperative projects led to what the editors of this volume call "investments in exploration"[1]: the creation of several major new research fields in the Netherlands. One interesting example in the early 1960s is space science. A striking feature of all these projects was the role of Philips Electronics, one of the largest and most powerful Dutch companies. Philips did not always remain active in the new fields, but even

D. Baneke (✉)
Utrecht University, Utrecht, Netherlands

© The Editor(s) (if applicable) and The Author(s) 2016
T. Heinze, R. Münch (eds.), *Innovation in Science and Organizational Renewal*, DOI 10.1057/978-1-137-59420-4_7

if it pulled out, it had often contributed significantly to the establishment of a new research infrastructure.

By following space science in the Netherlands from its beginning until the 1980s, we can investigate the interplay between national, industrial, and academic considerations in the establishment of a new scientific and technological field, reconsidering, for example, the importance of political considerations and commercial constraints, the role of management cultures, and the adaptation of institutions to changing contexts. This will enrich our understanding of the various roles that academic science, industrial companies, and the government (the three sectors that together form the so-called "triple helix," although that notion has been specifically applied for a more recent kind of cooperation)[2] played in the science infrastructure. These roles were not always clearly delineated.

In a 2006 paper, Philip Scranton called for a richer understanding of the role of non-market (government) actors in defining problem sets for innovation in the post-war period.[3] Scranton focused mainly on national security issues during the Cold War. The "military-industrial complex" of that era, or comparable networks of industry, academia, and government institutions, was a model of institutional cooperation in innovation and development between the three "triple helix" sectors in the post-war decades. Different models existed as well, however. Unlike in the USA, Britain, France, or Sweden, the military played only a small role in Dutch big science projects. Industry did, with an especially central role for Philips Electronics.

This chapter starts with an introduction of Philips Electronics and Fokker Aircraft and their roles in the Dutch national innovation system. Then I will describe their involvement in the establishment of a Dutch space program, focusing on the Astronomical Netherlands Satellite (ANS) project. Interestingly, the two companies drew different lessons from the project. I will analyze this difference by comparing their aims and ambitions, internal organizations, and the place of technological capability and innovation in the corporate identity of either firm. If we want to understand why Philips was such an important node in the scientific infrastructure, we have to realize that the boundaries between commercial, scientific, and national security considerations were not clear-cut. Philips was a commercial firm, but it also had internalized roles that are traditionally assigned to government or academia. Different parts of the company cooperated in almost the same way as university laboratories and industrial companies would. The case of Fokker was different. There, the notion of a "development pair" would be more applicable. A recent book about Swedish technology development used this notion for the close cooperation between a private company and a government institution in high-tech development projects, in which the government funds a private development project and acts as a guaranteed first buyer. A special version is an "auxiliary development pair," in which the government's support is not aimed at developing and procuring a specific product, but at indirectly supporting whole industrial sectors to build up their institutional infrastructure.[4]

The chapter ends with a brief discussion of the second Dutch satellite project, the Infrared Astronomical Satellite (IRAS), and the new innovation policy that emerged in the 1980s—a policy that was aimed at creating an innovation infrastructure that resembles what later became known as "mode 2" knowledge production.[5] As we shall see, this policy stimulated some forms of cooperation but terminated others. Notions such as "military-industrial complex," "non-military academic-industrial complex," "development pair," and "auxiliary development pair" all describe models of cooperation between governments, industry, and universities that predate the oft-discussed mode-2, but they do not resemble in any way the "mode 1" knowledge production as it is often summarily described in the mode-2 literature. As others have observed before, mode-1, like the "ivory tower" university or the "linear model" of scientific innovation, never existed except as an idealized model to clarify its opposite.[6]

The case of space science is especially interesting because the field did not just pose scientific and technological challenges but also organizational ones. For the emergence of space science as a new discipline, institutional and management innovation was as important as scientific and technological innovation. New forms of management knowledge had to be imported, in this case especially from? the National Aeronautics and Space Administration (NASA). Space projects were notoriously complicated, not only because of the extreme demands on quality and precision but also because of the number and variety of institutions that were involved. They were training areas for cooperation between scientists, engineers, business leaders, and government officials, and in many cases military officers as well.[7] Especially Fokker considered it crucial to learn how to manage large technological development projects, in other words: how to manage technological innovation. The accompanying management jargon formed a major part of the new communal language that all the actors in the new research field of space science had to master in order to be able to cooperate.[8]

Innovation was a central feature of the corporate identity of both Philips and Fokker. It was a part of their role as national champions and arsenals of knowledge and skill. For several reasons, building a scientific satellite was an excellent means to develop desirable skills. The "pure" scientific research that was done with the satellite was almost a by-product of the technology, not the main goal—a spin-off, so to speak. But in the long run, the science was perhaps the most important outcome. Especially IRAS produced ground-breaking new knowledge. This is a reversal of the standard narrative about the relation between science and technological applications.[9] It provides an interesting perspective on one theme of this book: the relation between institutional and intellectual change.

7.2   THE ROLE OF PHILIPS IN POST-WAR DUTCH SCIENCE

In the 1950s and 1960s, Philips Electronics produced a wide range of products, including of course lighting, but also domestic appliances, medical systems, and scientific instruments.[10] The company was constantly expanding. In the early 1970s, at the height of its power, Philips had more than 400,000 employees, including nearly 100,000 in the Netherlands (population at that time: 13 million). In the Netherlands, it was by far the largest company in its sector. Philips was considered to be a national champion: by the government, by the public, and also by itself. Traditionally, the company supported a wide range of social and cultural projects in the Netherlands. Especially in the Eindhoven area, Philips was omnipresent in housing and health projects, sports, and many other aspects of society (*Philips Sport Vereniging*, PSV, is still one of the major soccer teams of the country). These activities strengthened the company's standing as a national institution.

Engineering capability featured prominently in Philips' self-image. The firm's motto in the 1950s was "Triumph of Technology" (*Triomf der Techniek*). Obtaining technological knowledge in new fields was thought to reinforce the company in more ways than just future profitability. It was closely related to national political concerns about the (presumed) Dutch technological lag behind leading nations after the Second World War, and the ambition to maintain national capability in strategic fields. This "arsenal of knowledge" argument played an important role in national industrial policy.[11] Although Philips was not directly supported by the government, at least not openly, the firm always maintained close relations with the Economic Affairs ministry in The Hague.

The company invested heavily in research, spending up to 6% of the turnover on research and development in the 1950s. This money went to the development laboratories attached to each product division, but 1% of the turnover went to the *Natuurkundig Laboratorium* (Physics Laboratory, usually known as NatLab), an independent entity within the company.[12] NatLab founder Gilles Holst and his successor Hendrik Casimir were proud to make their laboratory an academic-style institution which spent significant sums on fundamental research.[13]

Together with Royal Dutch/Shell, by far the largest Dutch (or rather Dutch-British) company, Philips was the largest employer of physicists and chemists in the Netherlands. Recruiting talented students was a prime concern for the two multinationals. For that reason, they carefully cultivated their connections to universities. Several Philips scientists, including the directors of NatLab, had part-time professorships in Leiden or Delft, and academic professors regularly lectured at NatLab seminars.[14] Philips and Shell were important actors in the national research infrastructure. In the 1930s, they lobbied to establish graduate programs in Applied Physics at various universities. In the Interwar years, up to one third of the physics PhDs found jobs at those two companies.[15] The physics students of the Free University of Amsterdam even composed a special hymn for graduates who obtained a job at Philips.[16] After the Second World War, Philips and Shell donated large sums of money toward the founding of new laboratories and technology institutes. Philips was also represented in the governing boards of several universities. The exchange of staff between the universities and the industrial laboratories increased as well. According to Baggen, Faber, and Homburg, the companies significantly influenced academic research topics.[17]

An important aspect of Philips' corporate philosophy was that the company had to be involved in all major new fields of science, regardless of short-term expectations of profit or practical use. Board members Frits Philips and Th. Tromp considered cultivating a broad in-house scientific and technological capability to be crucial for the future of the company.[18] It would put the company in a position to quickly understand new developments and react to them if necessary. One never knew which technology would be the "next big thing," so one needed to have an arsenal of knowledge to draw upon. For this reason, novelty by itself was a motivation to invest in a new field, regardless of its immediate usage perspective.

Another reason was to make the NatLab an appealing employer for talented students with scientific ambitions.[19]

According to Casimir, the best way to get involved in scientific research was to develop scientific instrumentation.[20] One prime example was the production of electron microscopes, but Casimir was also interested in semiconductors and superconductivity, for example. After the Second World War, Philips became involved in new research organizations for, among other subjects, nuclear science and computing. In all cases, it provided instrumentation, most famously a cyclotron. It also got involved in uranium enrichment, first through a research institution and later as a stakeholder in the company Urenco.[21] Another new post-war research organizations was devoted to radio astronomy. During the war, Dutch astronomers Jan Oort and Henk van de Hulst had made plans for postwar radio astronomical research. After the war, Philips joined Leiden and Utrecht observatories in founding the Foundation for Radio Astronomy (SRZM). Over the next few decades, Philips supplied receivers and other technology for several radio telescopes. The combined interests of industry and scientists had enabled the foundation of a new field.[22]

The scope of Philips' activities made the company an important node in the national innovation system. In the context of government–industry–university relations, treating Philips simply as industry would be a mistake in this period. It had internalized elements of all actors. No other company had a comparable position. The embodiment of its scientific clout was NatLab's director H.B.G. Casimir, the most prominent Dutch physicist after the Second World War, who later became President of the Royal Netherlands Academy of Sciences. Another notable Philips alumnus was C.J. Bakker, who was involved in Philips' cyclotron project and later became director-general of the European Organization for Nuclear Research (CERN).

7.3   Fokker: The Importance of Being Creative

Fokker was not a member of the select group of companies that were responsible for the lion's share of R&D spending in the Netherlands.[23] It was, however, the only Dutch aircraft manufacturer, which made it a flagship company with a high national profile. It was a matter of national policy that the Netherlands should retain an independent and "creative" (zelfscheppende) aircraft industry, meaning that it should have the capacity to design, develop, and produce new aircraft models.[24] In order to support and fund this capacity, the government founded the Dutch Institute for Aircraft Development (NIV) in 1947.[25] In theory, Fokker would repay the cost of development projects to the NIV out of the profits made from those projects. Those funds could then be used for new projects, making the NIV a so-called "revolving fund." In practice, however, this rarely happened. Government funding for the NIV was a subsidy rather than an investment.

Because of its "creative" identity, engineering capability was at least as important for Fokker as for Philips, but other than for Philips, scientific novelty was less important than technological independence. Its focus was on development rather than research. Fokker had no academic-style laboratory, nor did the company try to be involved in all new high-tech fields. It had a well-defined core business, which it strove to strengthen by technological innovation. At the same time, its national flagship status and the government support through the NIV sometimes clashed with commercial considerations. This sometimes caused tensions in the company's management. For example, it was understood that Fokker needed foreign partners to survive commercially, but its (government-backed) insistence on an independent Dutch engineering department made it difficult to cooperate successfully. Joint ventures with the German Vereinigte Flugtechnische Werke (VFW) as well as with McDonnell Douglass, Aerospatiale, and British Aerospace proved unsuccessful in the long run, in no small part for that reason.[26]

The relationship between Fokker and the NIV could be viewed as a "development pair," except that in the case of Fokker, the Dutch government could not guarantee to act as a first buyer of the end products of the joint development projects. The national airline KLM was independent enough to purchase other aircraft models if it wanted (which it often did), and the Defense ministry often chose not to buy Fokker models that were adapted for military use.[27] This obviously caused some frustration at Fokker.

7.4   Establishing a New Field: Space Research

Scientific research with instruments outside the Earth's atmosphere started after the Second World War. In the USA, the Soviet Union, and (on a smaller scale) in France and Britain, captured German V2 rockets were used to observe, for example, the earth's magnetic field and cosmic radiation. These experiments were difficult, yielding only a few minutes of observations per flight, with a high failure rate. For most scientists, systematic space research became a realistic possibility only after 1957, when Sputnik proved that a longer stay in space on a relatively stable platform was possible.

Satellites were, of course, dependent on military rocket (missile) technology to put them into orbit. In the 1950s, only the USA and the Soviet Union possessed this technology. For scientists from other countries, space came (literally) within reach after American diplomats announced an offer to launch foreign scientific experiments on American rockets. This was a part of their strategy to emphasize their openness and peaceful intentions, in contrast to the secretive Soviet Union.[28]

In the Netherlands, discussions about a national space program started in 1959, when Dutch minister of Foreign Affairs (and future Secretary General of NATO) Joseph Luns wondered how the Netherlands could get involved in space activities. Luns stated that for political, scientific, technological, and commercial reasons, the Netherlands could not afford to be left behind. The Dutch ambassador in the USA had already warned that NASA officials, who were looking for foreign partners, had gotten the impression that there was no relevant Dutch institution to talk to. The ambassador had pointed specifically at the opportunities that space activities offered for Fokker and in the field of "electronics."[29] Among the first to react to Luns' inquiries were the astronomers Jan Oort and Henk van de Hulst. They had no experience with space research—both were active in radio astronomy—but Van de Hulst was president of Committee on Space Research (COSPAR), an international committee of scientists for the advancement of the scientific use of space technology. Around the same time, Eduardo Amaldi and Pierre Auger launched a plan for European cooperation in space, modeled after CERN. Van de Hulst was involved in the discussions about this plan because they took place in the margins of COSPAR meetings.

Luns wanted to join the European space effort, "both because of the countries that will join this European organization, and for financial, personal and scientific reasons."[30] He hoped that the new organization would cooperate with the USA, to benefit from America's technological prowess. Simply joining the talks was not enough, however: he wanted the Dutch opinion to carry weight in the negotiations. The best way to ensure influence would be to have a "modest but sophisticated" (*bescheiden maar weloverwogen*) domestic space program. Luns expected that the national science community and the flagship companies, with their arsenals of knowledge, would enable the Netherlands to enter this new field with relative ease. At the same time, those same institutions stood to be the main beneficiaries.

It took until 1964 before a European space organization was founded— or rather, two organizations: the European Space Research Organization (ESRO) and the European Launcher Development Organization (ELDO).[31] Until ELDO had produced its rocket, ESRO would make use of the American offer to launch foreign experiments. The Netherlands joined both organizations. The Dutch contribution to the ELDO launcher was coordinated by the Institute for Aircraft Development and the Dutch Aeronautical Laboratory (NLL). Both Fokker and Philips (especially its telecommunications division) participated in it. The Dutch participation in ESRO was coordinated by the Geophysics and Space Research Committee (GROC) of the Royal Academy of Sciences.[32] This committee was dominated by astronomers, with Van de Hulst acting as chairman. This rather informal, ad hoc organization coordinated Dutch space science until the mid-1980s.

During the next few decades, all major Dutch space research projects were astronomical experiments.[33] There are several reasons why the astronomers were able to monopolize the field. Most importantly, they had created a strong institutional infrastructure that enabled them to react quickly to new developments and to cooperate on a national level. They had both organizational experience and excellent contacts in political and industrial circles, including with Philips (via radio astronomy). Besides, the Dutch "school" of astronomy had an excellent international reputation.[34]

7.5 The Need for a Large National Project

Fokker, Philips, and the Dutch government had hoped to secure large development contracts from the new European space organizations, but after a few years it became clear that this would not happen. Both companies blamed their lack of proven experience in space projects, but also the fact that the contracts of these organizations were awarded proportionally to a nation's contribution, which in the Dutch case was relatively small. For that reason, the companies lobbied for a significant expansion of the national space program.[35] A large domestic project would provide them with experience and know-how, while at the same time offering the opportunity to demonstrate their capabilities to potential customers.

Acquiring new technical knowledge was not the main argument of the two companies. The "spin-off" effect of space technology for aircraft development was expected to be fairly small. The transfer of skills the other way around was expected to be much more significant: both Philips and Fokker expected to be able to enter the new field easily, cashing in on the arsenal of knowledge it had built since the war. One Fokker engineer called this "spin-in" instead of spin-off.[36]

Crucially, Philips and Fokker both argued that organizational knowledge and managerial experience were at least as important as technological innovation. This argument was used and repeated by industry lobbyists, ministry officials, and politicians alike.[37] In the 1960s, project management was regarded as crucial to innovation. "Systems Management" became a key modern technology in the era of big development projects that had to deal with many actors from various disciplines and institutions, large uncertainties, complex flows of information, and especially constantly changing objectives and design specifications. Developed by the US Air Force and aerospace industry, it was perfected in the Apollo project, generally hailed as a triumph of management as well as technology. The European space organizations ESRO and ELDO tried to emulate this success, with varying results. Especially ESRO looked at NASA as a model for project management.[38]

Obtaining and demonstrating the capability to manage complex development programs was especially important for Fokker. While Philips traditionally entered new fields by developing components or instruments, Fokker wanted to work on the highest "system" level. The emphasis on management skills was related to the national policy of maintaining a "creative" national industry, which attached much value to technical development activities. Fokker's space activities were not expected to be commercially profitable in the short or even medium term, but rather to support the company's (and by extension the nation's) corporate standards, project management skills, quality control, and morale.[39] The management techniques of space projects, with their emphasis on reliability, quality control, and integral system engineering, were directly applicable to aircraft development.

At Philips, similar arguments were used. It had a rather complex internal structure, with many semi-independent units, including national branches in several countries, specialized product divisions, and the NatLab. There were no multi-disciplinary development projects on the scale of Fokker's aircraft development. Still, Philips' Central Technical Efficiency and Organization department promoted systems management, attempting to streamline development and production efforts and to make the various departments cooperate more efficiently.[40] Besides, J.H. Spaa of Philips' Central Development Bureau argued that high-profile development projects boosted corporate confidence.[41] At the same time, as we shall see, market considerations and profitability played a more important role at Philips than at Fokker.

7.6   THE ANS

In response to the industrial lobby, the Dutch government sent a call for proposals for an extended national space program to industry, and also to the Royal Academy of Sciences. Fokker, Philips, and the astronomers carefully coordinated their answers. They all proposed to build an astronomical satellite, to be launched with one of the rockets that the American government had offered for foreign science instruments. This became the ANS.

Fokker and Philips mainly wanted to build a satellite; they did not much care about what it would be used for. For several reasons, an astronomical satellite perfectly matched their ambitions: the international prominence of Dutch astronomy justified a large public investment; the project would be unique; astronomy was easy to popularize, making the project visible; it would provide ample opportunity to exhibit technological skill; and finally it was not so politically complicated as, for example, communications satellites.[42] Another reason, not mentioned by the companies, might have been that since ESRO was the main potential client, it was important for industry to demonstrate that it could cooperate with scientists.

Both industry and the astronomers wanted the satellite to be eye-catching, the former because it wanted to advertise, the latter because they wanted to operate at the forefront of science. The satellite would get an innovative stabilization and pointing system, for example. For similar reasons, Philips provided an advanced reprogrammable on-board computer.[43] Fokker built the satellite frame. The scientific instruments were provided by the universities of Utrecht and Groningen. According to Utrecht astronomer De Jager, Philips and Fokker accepted the scientific instrument proposals without any discussion.[44] It was clear that for the companies, as for the government, science was not the main goal of the mission.

ANS was to be launched on an American Scout rocket. NASA even provided a slightly larger launch vehicle, in order to be able to add an American instrument to the mission, which was interpreted by the Dutch as a vote of confidence in the project.[45] But NASA also provided support in the form of rigorous reviews at moments of design "freezes," as well as advice on procedures for component specifications and quality assessment, and how to manage design changes on various levels. These standardized procedures were new to both Philips and Fokker. Additional support was provided (for a fee) by General Electric (GE). Besides, Fokker staff spent several months at GE and Republic Aviation (owned by Fairchild) to learn some aspects of space technology.[46]

ANS was launched in 1974. Due to a minor malfunction, its orbit was more elliptic than planned. Philips' eagerness to show off paid off in this case: much of the observation program could be saved by reprogramming the on-board computer. The science results were respectable but not spectacular; however, the technological performance of the satellite was excellent. The total cost of the mission was estimated to be close to ƒ100 million, almost twice the original estimate. Fokker and Philips reported that they invested ƒ13 million for design studies, fees for GE, and renounced profit.[47] Unofficially, Philips' estimated investment was higher (see below).

One could describe the relation between the government and industry in the ANS project as a "development pair." Most of the funding came from the ministry of Economic Affairs, with a smaller contribution of the ministry of Science and Education. The (government-funded) astronomical community was pushed forward as the first buyer of a space satellite, with the explicit intention of paving the way for future commercial customers. Of course, one has to remember that a satellite was not a new type of car or even jet fighter. Serial production would not be an option in space technology for many decades to come.

The conditions were specified in a contract, which included strict conditions about the price in case of delays or cost overruns. This type of government sponsoring by development contract was a novelty at the time, and it was expected that more would follow. For that reason, Spaa advised the Philips board that the company's contribution should not necessarily be large, but it should be highly visible, for political reasons. In a later stage, board member and NatLab director Casimir also argued that Philips should accept financial loss on this project in order to secure the government's goodwill for future projects.[48]

7.7   Lessons from ANS: Difference Between Philips and Fokker

Around the time of the launch of ANS, representatives of industry and astronomers discussed the possibility of an "ANS-B," a second Dutch satellite based on the same design, again with American cooperation.[49] The proposals referred to the same arguments about the importance of technical and managerial experience for industry and innovation. A new argument was added however: space technology neatly matched the new economic policy aims of the government in the 1970, because of its relatively small use of raw materials and energy and its potential application to monitor environmental problems.[50]

The new satellite was mainly promoted by Fokker. Philips supported the lobbying effort, but behind the scenes the company's management had already decided to pull out of the space business. This was the result of an internal evaluation of the ANS project. The remarkable difference between the two firms' evaluations reveals the different corporate strategies concerning innovation, which directly influenced their role in space science, the field that they had helped to build.

At Fokker, the space activities had been concentrated in a dedicated department, to which staff was allocated on a temporal basis as the project required. This matched the existing company structure: large, multi-disciplinary development projects with tight quality constraints were part of the normal way of operating in aircraft manufacturing. That is also why the company was so interested in NASA's project management procedures. ANS was a large and complex project in which every design change, no matter how small, had consequences throughout the system, which was exactly what made it so interesting to Fokker.

Things at Philips were different. The company had a venerable tradition in scientific research and high-tech development, but ANS was the first project of this magnitude.[51] Work on the project had been divided over several of the relatively independent units within the company. Much of the most innovative technical work was done by a relatively isolated group within the NatLab; the on-board computer was built by the subsidiary Hollandse Signaalapparaten, a defense contractor, while the Telecommunications division (*hoofdindustriegroep* PTI) provided components, as did other divisions. This complex institutional structure had impacted the project in several ways. Philips was a microcosm, in which various features and problems of university–industry cooperation were visible. Some divisions complained that weight and power allowances within the satellite were not distributed fairly between the components, making the margins for their work extra tight. At NatLab, staff complained that its mission was to do research, not coordinate large-scale projects. Apparently, the interest in management was stronger at the central company level than in the NatLab or the divisions. Meanwhile, the telecommunications division complained that it had been left with relatively uninteresting but costly work. The components themselves were not so novel as to require innovations that could be used in other products, while the quality constraints were a thousand times stricter than the division was used to. The division was compensated for this work—it was treated as a subcontractor—but still, manager N. Rodenburg was very worried about the financial consequences of the project.[52] In an evaluation of the management aspects of ANS, Philips engineer P. van Otterloo concluded that the complexity of ANS had been underestimated.[53] Project planning procedures had struggled to keep up with the frequent design changes, while paperwork and quality assessment had cost much more time than expected, resulting in delays and cost overruns. As the government contracts specified a fixed price with only a partial reimbursement of budget overruns, ANS left Philips with an estimated loss of c. ƒ17.5 million.[54]

Despite these problems, Van Otterloo considered ANS a useful project for Philips, not least because it was a "valuable exercise in the application of Systems Management in a Research and Development project."[55] During the project, the company's staff had learned the new language of component specifications, systems design reviews, failure mode and effect analysis, and other management procedures. These notions were increasingly regarded as useful tools in both development and production. Van Otterloo suggested that ANS could be a useful case study for the company's training program for talented young staff members for this reason.

An independent consultant, General Technology Systems Ltd, also concluded that the fragmented internal organization negatively impacted the firm's prospects in space activities. For example, the isolated position of the ANS project group at NatLab made it hard for other Philips departments to benefit from the gained technical knowledge.[56] In the end, Philips' leadership concluded that it had no future in space. Only Hollandse Signaalapparaten would remain active in the field. The project had been an interesting technological challenge, but the multi-disciplinary aspect of the project was not very interesting to the company, especially compared to the huge administrative effort and the amount of staff and resources that had been invested. Space projects were too complex and too unpredictable, and they did not fit the company's structure.[57] Besides, Philips was increasingly skeptical about the commercial outlook for space products. The international market was difficult to penetrate, while the national market was simply too small. Similar reasons had led Philips to abandon its ambitions in the field of nuclear energy.

Fokker's role more traditionally matched that of industry, though it was shielded from direct market pressure by direct and indirect government support. Other than Philips, Fokker had no ambition to do academic-style scientific research, though it was keen on producing new knowledge, both in technology and in management. Fokker was not put off by bureaucratic complexity and extreme quality constraints. Learning how to manage those was crucial for its core business. Nor was it deterred by commercial uncertainty, as that too was common in the aircraft business. The company's monolithic structure made it relatively easy to allocate staff to temporary programs within the company. Besides, the semi-public NIV bore most of the financial risks of its development projects. Just as with aircraft development projects, Fokker promised to repay the NIV's investment with profits obtained from future contracts, but in the case of ANS no one really expected any profit in the short or even medium term.[58] Fokker got exactly what it wanted out of the project—except international contracts, which was why it wanted to build another, more ambitious, national satellite.

7.8   IRAS and the Policy Changes in the 1980s

Both Fokker and the astronomers were pleased with the ANS project, and eager to initiate a second project along similar lines. Although they were disappointed about Philips' decision to terminate its space activities, they obtained Philips' promise to politically support a campaign for a second scientific satellite.[59] The campaign was successful: the government agreed to a second national satellite, again mostly funded by the ministry of Economic Affairs. This became the IRAS.

IRAS was a much more ambitious project than ANS, not least because it included cryogenic cooling of the complete telescope system. It became even more complicated when NASA decided to merge it with several American proposals for infrared satellites. IRAS became a joint American–Dutch project, with the Americans supplying crucial technology and half of the total funding. Great Britain also participated in the project, providing the ground station. Throughout the project, IRAS was plagued by problems, both technologically and organizationally.[60] The satellite was launched in 1983. It provided the first infrared survey of the sky, including observations of interstellar dust clouds and thousands of new objects. The IRAS catalogues of observations became starting points for much subsequent astronomical research. The cryogenic technology was later used in several other satellites, including the Cosmic Background Explorer which earned its principal researchers a Nobel Prize, and Gravity Probe B.

Of course, both Fokker and the astronomers were eager to build a third satellite. This time, the astronomers proposed an X-ray observatory. An influential government council advised negatively, however. The Dutch government had funded ANS and IRAS to help Dutch industry to enter a new market; it was about time that the space sector should become economically independent. But it was not only reluctance to keep funding one economic sector that withheld the government. More generally, views on the government's role in industry and innovation had changed. Politicians had become wary of directly subsidizing large industries after the messy bankruptcy of the Rijn-Schelde-Verolme (RSV) shipyards in 1983. Besides, there were increasing European regulations against state support for industry. Finally, changing views on market (de)regulation also worked against supporting individual companies. In the political and economic context of the 1980s, direct government support for large companies was not as natural as it had been before, although both Fokker and Philips kept receiving support behind the scenes (e.g. with the controversial "Technolease" construction).

For these and other reasons, government policy changed from targeted support to a more general "innovation policy," which explicitly would also include small and medium-sized companies. The new aim was to stimulate market-driven cooperation between industry and academia, preferably without too much government interference or funding. The government attempted to do this by creating favorable institutional frameworks and incentives.[61] This meant that space science and nuclear science, two of the main beneficiaries of post-war science policy, lost their privileged position.

Together with other developments at universities and in industry, the new policy opened the way for the emergence of what is often described as "mode 2" knowledge production frameworks, or something closely related.[62] At the same time, this meant the end of the kind of cooperation that produced ANS and IRAS. As we have seen, this was as much the result of changes in economic policy as in innovation or science policy. The immediate result was that there would be no third national satellite. Henceforth, all space activities would take place in the context of NASA and European Space Agency (ESA) missions, "as befits a small nation," in the words of Science minister A. Pais.[63] This was both because the cost of space missions had increased and because after years of struggling, ESA had finally become a successful organization with its own launch capability (the Ariane launchers).

The Netherlands no longer tried to maintain an independent capacity to build entire satellites, but rather specialized in specific components.

The changing political climate also had direct consequences for the institutional organization of space research. The informal structure of the Royal Academy Committee on Space Science (GROC) was replaced by a more formal organization, modeled after the existing organizations for nuclear physics and radio astronomy. One could say that, space science became a "normal" scientific discipline. The new Space Research Organization (SRON) was still funded by the government, but it was also supposed to earn 15% of its budget by doing contract research for industry. This is an example of way the government tried to press institutions to enter new partnerships. The government suggested that its skills in high-precision manufacturing, miniaturization, and robotics might be useful for medical appliances, for example. In practice, this proved to be difficult. The largest contracts came from science organizations such as ESA and CERN, all government-funded organizations.[64]

The changed socio-economic context also had consequences for Philips and Fokker. They felt the increased market pressure, but again, they chose radically different solutions. Philips finally gave up its ambition to maintain a complete arsenal of knowledge. In a series of radical reorganizations, the company terminated or scaled down its activities in many fields, focusing on a number of core areas such as lightning and medical systems. The number of staff also decreased significantly. In the best-known reorganization, "operation Centurion" in the early 1990s, the complex structure of the firm was streamlined, reducing the number of divisions and departments. One could perhaps say that financial and commercial pressure forced a change in emphasis from engineering to commerce. The NatLab was also downsized and its "pure science" ambitions were toned down, although it remained by far the largest industrial laboratory of the Netherlands. Philips focused more on its role as a commercial industrial firm and less on the academic and national warehouse of knowledge aspects. So ironically, in the era of increasingly dynamic cooperation between industry, research institutions, and government organizations, some types of crossovers ended.

Fokker chose an opposite approach. Its focus on engineering and large-scale development increased rather than decreased. Fokker's space department finally managed to obtain several contracts, usually as part of international consortia. After IRAS, it did not build complete spacecraft, but gradually specialized in components such as solar panels. In the mid-1980s,

Fokker also started two major new aircraft development projects (F50 and F100). These projects proved to be too ambitious, however. The company became increasingly dependent on government subsidies. Foreign partners were sought, but as before, this was complicated by the Dutch insistence of maintaining an independent engineering unit in the Netherlands.[65] In 1996, Fokker had to file for bankruptcy. The space department survived, as it had become independent company shortly before the bankruptcy. Under the name Dutch Space, it is now part of Airbus Defense and Space, a European aerospace company.

7.9   CONCLUSION

The establishment of space research as an academic research field in the Netherlands was the result of a complex mixture of political, economic, scientific, and institutional developments. It was the Foreign Ministry that first raised the subject, but Philips and Fokker were the driving forces behind the Dutch national space program in the 1960s and 1970s. Their political clout provided astronomers with some of the most expensive scientific instruments ever built in the Netherlands. Astronomy benefited as vehicle for government support as "first buyer," in an institutional setup that resembled a "development pair."

ANS and IRAS were scientific instruments, used by the traditional academic discipline of astronomy. They became the flagship projects of a new research field: space research. But big science is never just about science.[66] The case of Fokker illustrates the importance of development rather than research. It also illustrates that companies do not need to do fundamental science to have a major impact on the development on a scientific field.

Many arguments were used to legitimize government spending on space technology. Significantly, the introduction of innovative management systems was one of them. Scranton has stressed the importance of management techniques in post-war innovation.[67] Cold War era development projects were so complex and unpredictable that cost and risk management was extremely difficult. Controlling them became a key technology in itself. In this case, a demand for institutional renewal motivated the establishment of a new research field as much as the other way around!

The importance of management skills also illustrates that universities or industrial research laboratories are not the only source of innovation. Important types of new knowledge were produced at other levels. Focusing on academic-style research as the main source of new knowledge misses important aspects of innovation. Similarly, the arguments for cooperating with NASA show that importing knowledge was as much a source of new skills as in-house innovation. This goes especially for institutional innovation.[68]

Philips' unrivaled position in the Dutch economic and scientific landscape was crucial for the formation of several new research fields. Even when the company was not able to gain a strong position in a new market, its efforts had a lasting impact on the Dutch scientific infrastructure, and thus to the renewal of Dutch science (see editor's introduction). Few technological companies had a similar broad and deep presence in their home country. The most comparable case might be Sweden, where SAAB and other Wallenberg group industries also acted as national institutions as well as commercial firms. The relation between Philips and academic institutions was so systematic that it can be compared to Eisenhower's military-industrial complex, except that in this case the military were not involved.

Scranton has mentioned several ways in which governments can stimulate industrial innovation: by stimulating innovation in state-owned firms or by initiating "projects" in cooperation with industry.[69] Other models include cooperation in a "development pair," large-scale government (military-) industrial "complex," or governments acting as a guaranteed first buyer of an innovative product. Governments, private companies, and research institutions were involved in ever-changing institutional setups throughout the twentieth century (and probably also before). The view of science as a "source of strategic opportunity," one of the characteristics of "mode 2" knowledge production, is by no means recent or new.[70] The history of innovation since the Second World War is much richer.

Only in the late 1970s did the government start to develop an innovation policy. The idea itself was not new; the novel aspect was the fact that it was an explicit policy instead of a seris of ad hoc decisions. This gave rise to new tools and concepts. ANS was never part of an "innovation policy"; it was industrial policy and science policy. When this specific kind of industrial policy fell out of favor in 1980s, this led to the cancellation of a third national satellite. The emergence of mode-2 as model favored by policy makers meant the end of some other models. Interestingly, science policy since the 1980s has been increasingly aimed at using science to support innovative industry. In this case, however, the opposite happened: industrial policy supported the emergence of a new scientific field. This was not the main goal, but it was perhaps the most notable effect.

NOTES

1. See editor's introduction, in this volume.
2. Henry Etzkowitz and Loet Leydesdorff, "The Dynamics of Innovation: from National Systems and 'Mode 2' to a Triple Helix of University-Industry-Government Relations," *Research Policy* 29 (2000); Henry Etzkowitz, "Innovation in Innovation: the Triple Helix of University-Industry-Government Relations," *Social Science Information* 42 (2003).
3. Philip Scranton, "Technology, Science and American Innovation," *Business History* 48 (2006): third proposition.
4. Per Lundin, Niklas Stenlås and Johan Gribbe, *Science for Welfare and Warfare: Technology and State Initiative in Cold War Sweden* (Sagamore Beach, MA: Science History Publications, 2010), 45, 147, 255.
5. Michael Gibbons et al., *The New Production of Knowledge: the Dynamics of Science and Research in Contemporary Societies* (London: Sage, 1994); Helga Nowotny, Peter Scott, and Michael Gibbons, "Introduction: 'Mode 2' Revisited: the New Production of Knowledge," *Minerva* 41 (2003); Dominique Pestre, "Regimes of Knowledge Production in Society: Towards a More Political and Social Reading," *Minerva* 41 (2003).
6. Pestre, "Regimes of Knowledge"; cf. David Edgerton, "The 'linear model' did not exist: Reflections on the history and historiography of science and research in industry in the twentieth century," in *The Science-Industry Nexus: History, Policy, Implications*, ed. Karl Grandin and Nina Wormbs (New York: Watson, 2004).
7. Stephen Johnson, *The secret of Apollo: systems management in the American and European space programs* (Baltimore: Johns Hopkins University Press, 2002).
8. Space research can be regarded as a 'fractionated trading zone' as described by H. Collins, R. Evans, and M. Gorman, "Trading zones and interactional expertise," *Studies in History and Philosophy of Science* 38 (2007): 660–62, in which various disciplinary cultures remained visible next to each other while sharing important material cultures (the projects were centered on space instruments) and a communal language (systems management).
9. Cf. Scranton, "American Innovation," second proposition.
10. On the history of Philips after the Second World War: I.J. Blanken, *Een industriële wereldfederatie: Geschiedenis van Koninklijke Philips Electronics N.V., vol. V* (Zaltbommel: Europese Bibliotheek, 2002).
11. On the notion of "arsenal of knowledge": John Krige, "Building the arsenal of knowledge," *Centaurus* 52, no. 4 (2010).
12. Blanken, *Een industriële wereldfederatie*, 129, 147.

13. Kees Boersma, *Inventing Structures for Industrial Research: a History of the Philips NatLab 1914–1946* (Amsterdam: Aksant, 2002); Marc de Vries, *80 years of research at the Philips Natuurkundig Laboratorium 1914–1994* (Amsterdam: Pallas Publications, 2005).

14. Marijn Hollestelle, *Paul Ehrenfest: Worstelingen met de Moderne Wetenschap, 1912–1933* (Leiden: Leiden University Press, 2011), 186.

15. H.G. Heijmans, *Wetenschap tussen universiteit en industrie: de experimentele natuurkunde in Utrecht onder W.H. Julius en L.S. Ornstein 1896–1940* (Rotterdam: Erasmus Publishing, 1994), esp. 160–161; cf. P. Baggen, J. Faber and E. Homburg, "Opkomst van een kennismaatschappij," in *Techniek in Nederland in de twintigste eeuw VII: Techniek en modernisering: balans van de twintigste eeuw*, ed. by J.W. Schot et al. (Zutphen: Walburg Pers, 2003).

16. A. Flipse, " 'Geen weelde, maar een offer'. De band tussen Vrije Universiteit en achterban, 1880–1950," in *Universiteit, Publiek en Politiek*, ed. by L.J. Dorsman and P.J. Knegtmans (Hilversum: Verloren, 2012).

17. Baggen, Faber and Homburg, "Opkomst".

18. Blanken, *Een industriële wereldfederatie*, esp. Chap. 4.

19. Ibid., 133.

20. De Vries, *80 years of research*, 234.

21. Friso Hoeneveld (Utrecht University) is working on a PhD dissertation on the history of the nuclear science institution (FOM). Abel Streefland (Leiden University) is working on a PhD dissertation on the Dutch uranium enrichment project. Albert Kersten, *Een organisatie van en voor onderzoekers: ZWO 1947–1988* (Assen: Van Gorcum, 1996); F. Hoeneveld and J. van Dongen, "Out of a clear blue sky? FOM, the bomb and the boost in Dutch physics funding after World War II," *Centaurus* 55 (2013).

22. On radio astronomy: Woodruff T. Sullivan, *Cosmic noise: a history of early radio astronomy* (Cambridge: Cambridge University Press, 2009); Astrid Elbers, "The establishment of the new field of radio astronomy in the postwar Netherlands: a search for allies and funding," *Centaurus* 54 (2012).

23. For most of the twentieth century, those companies were Philips (electronics), Shell (oil), Unilever (food and consumer products), AKZO and its predecessor AKU (chemical industry), and DSM (mining, later chemical industry). Currently, the largest R&D investors also include two former Philips subsidiaries, ASML and NXP (both semiconductor industry).

24. On Fokker: Marc Dierikx, *Uit de lucht gegrepen: Fokker als Nederlandse droom, 1945–1996* (Amsterdam: Boom, 2004); A.A.M. Deterink et al., *Onderzoek naar de oorzaak van het faillissement van Fokker* (Deventer: Kluwer, 1997).

25. Ed Muller, *50 jaar Nederlands Instituut voor Vliegtuigontwikkeling en Ruimtevaart* (Katwijk aan Zee: Satellite Services, 1997).

26. Deterink et al., *Onderzoek*.
27. Dierikx, *Fokker als Nederlandse droom*.
28. John Krige, *American Hegemony and the Postwar Reconstruction of Science in Europe* (Cambridge, MA: MIT Press, 2006a); John Krige, "Technology, foreign policy, and international cooperation in space," in *Critical issues in the history of spaceflight*, ed. Steven J. Dick and Roger D. Launius (Washington, DC: NASA, 2006b).
29. GROC file 165/347-16: J.C. Kruisheer to J. Luns, 10 May 1960; Henk van de Hulst, "Seizing opportunities: some comments on the Dutch national space science programme of the sixties and seventies," in *Science beyond the atmosphere: the history of space research in Europe*, ed. Arturo Russo (Noordwijk: ESA, 1992). On the history of Dutch space research: Niek de Kort, *Ruimteonderzoek: de horizon voorbij* (Amsterdam: Natuur & Techniek, 2003); David Baneke, "Space for ambitions: the Dutch space program in changing European and transatlantic contexts," *Minerva* 52 (2014).
30. Letter from Luns, 23 January 1960, National Archives, The Hague, Algemene Zaken records, file 5714.
31. John Krige and Arturo Russo, *A history of the European Space Agency 1958–1987, vol. I: the story of ESRO and ELDO 1958–1973* (Noordwijk: ESA, 2000).
32. Klaas van Berkel, *De stem van de wetenschap: geschiedenis van de Koninklijke Nederlandse Akademie van Wetenschappen, vol. 2* (Amsterdam: Uitgeverij Bert Bakker, 2011), 328–37.
33. A list of experiments are provided by De Kort, *Ruimteonderzoek*, 206.
34. David Baneke, "Teach and travel: Leiden Observatory and the renaissance of Dutch astronomy in the interwar years," *Journal for the History of Astronomy* xli (2010).
35. NA, Binnenlandse Zaken records, file 5577: letters from industry.
36. Interview by the author with Jan de Koomen (26 April 2011); cf. file 821:921.94 no.1b: "Some considerations on a scientific satellite", July 1963; cf Krige and Russo, *European Space Agency*, 73; J.H. Spaa, "Enige konklusies uit het ANS-projekt vanuit het Philips standpunt," Philips Company Archives, file 821:921.94 no. 4, 1975.
37. See, for example, NA, Binnenlandse Zaken records, file 5577; remarks in Parliament by Minister Nelissen (Economic Affairs) 12 November 1970, Handelingen van de Tweede Kamer 1970–1971, pp. 940–941; Jaaradvies RAWB 1976, Handelingen van de Tweede Kamer 1975–1976 document no. 13918 p. 28.
38. Johnson, *The secret of Apollo*.
39. Interviews by the author with Reinder van Duinen (26 August 2010) and Jan de Koomen (26 April 2011).

40. P. van Otterloo, "Management aspecten van het ANS project," Philips Company Archives, file 821:921.94 no. 4, 1973, 11.
41. Spaa, "Enige konklusies".
42. PCA file 821:921.94 no. 1, *Voorstel van de Nederlandse electronische- en vliegtuigindustrie voor de ontwikkeling van een Nederlandse astronomische satelliet* (1966).
43. De Vries, *80 years of research*, 234–37.
44. Kees de Jager, "ANS, de eerste Nederlandse satelliet," *Zenit* (2009); interview by the author with Kees de Jager, 7 April 2011.
45. For example, in a memo to the prime minister, 24 June 1976, NA, Algemene Zaken records, file 10110. See also NA, Binnenlandse Zaken records, file 5591.
46. PCA, file 821:921.94 no. 1, *Voorstel van de Nederlandse electronische- en vliegtuigindustrie voor de ontwikkeling van een Nederlandse astronomische satelliet* (1966); interview De Koomen.
47. Handelingen van de Tweede Kamer 1973–1974 document no. 12932. Muller, *50 jaar Nederlands Institut*, 86, estimated the total costs at ƒ150M, possibly correcting for inflation.
48. PCA file 821:921.94 no. 2, meeting report, 4 December1970.
49. Various correspondence about this in PCA file 821:921.94 no. 3; see also GROC file 347–6, minutes of the meeting of 28 September 1973.
50. For example: report *Ruimtevaart en nationale doelstellingen*, 1976, NA, Algemene Zaken Records, File 10110.
51. PCA file 821:921.94 no. 4: K. Woensdrecht to Pannenborg, 28 March 1974.
52. Ibid. no. 3: letter by N. Rodenburg to Casimir and Pannenborg.
53. Van Otterloo, "ANS project".
54. PCA file 821:921.94 no. 4: memo "totaal verlies van Philips aan de ANS", 25 March 1975.
55. Van Otterloo, "ANS project".
56. PCA file 821:921.94 no. 4: Final Report on a Study of ANS Benefits by General Technology Systems Ltd, 1977.
57. Ibid. no. 3: reports of a meeting on 3 March 1971.
58. Dierikx, *Fokker als Nederlandse droom*, 171–172.
59. Blanken, *Een industriële wereldfederatie*, 97–98; PCA 821:921.94 no. 3 and 4.
60. On IRAS: Mitchell Waldrop, "Infrared Astronomy Satellite," *Science*, 220 no. 4604 (1983); Wallace Tucker and Karen Tucker, *The Cosmic Inquirers: modern telescopes and their makers* (Cambridge, MA: Harvard University Press, 1986); Baneke, "Space for ambitions".

61. Ton van Helvoort, *De KNAW tussen wetenschap en politiek: de positive van de scheikunde in de Akademie in naoorlogs Nederland* (Amsterdam: KNAW, 2005); Van Berkel, *De stem van de wetenschap.*
62. See also Harry de Boer and Jeroen Huisman, "The New Public Management in Dutch Universities," in *Towards a New Model of Governance for Universities?*, ed. D. Braun and F.X. Merrien (London: Jessica Kingsley Publishers, 1999).
63. Letter from Minister A. Pais, 5 November 1980, GROC file 347-9/10.
64. Annual reports of SRON.
65. Deterink et al., *Onderzoek.*
66. Robert W. Smith, *The space telescope: a study of NASA, science, technology, and politics* (Cambridge and New York: Cambridge University Press, 1989); Peter Galison and Bruce Hevly, ed., *Big science: the growth of large-scale research* (Stanford, CA: Stanford University Press, 1992); Cooper H. Langford, and Martha Whitney Langford, "The evolution of rules for access to megascience research environments viewed from Canadian experience," *Research Policy* 29 (2000).
67. Scranton, "American Innovation," 322.
68. Cf. Bent Dalum, Björn Johnson and Bengt-Åke Lundvall, "Public Policy in the Learning Society," in *National Systems of Innovation: towards a theory of innovation and interactive learning*, ed. Bengt-Åke Lundvall (London and New York: Pinter, 1992).
69. Scranton, "American Innovation," 321.
70. Cf. Olle Edqvist, "Layered Science and Science Policies," *Minerva* 41 (2003).

Acknowledgments This chapter is based on research that was supported by a Guggenheim Fellowship at the National Air and Space Museum in Washington, DC and an NWO grant at the VU University of Amsterdam. I thank David DeVorkin, Frans van Lunteren, Harm Habing, and the editors of this volume for their helpful comments and suggestions.

REFERENCES

Baggen, P., J. Faber, and E. Homburg. 2003. Opkomst van een kennismaatschappij. In *Techniek in Nederland in de twintigste eeuw VII: Techniek en modernisering: balans van de twintigste eeuw*, ed. J.W. Schot, H.W. Lintsen, A. Rip, and A.A. Albert de la Bruhèze, 141–173. Zutphen: Walburg Pers.
Baneke, David. 2010. Teach and travel: Leiden Observatory and the renaissance of Dutch astronomy in the interwar years. *Journal for the History of Astronomy* xli: 167–198.

Baneke, David. 2014. Space for ambitions: The Dutch space program in changing European and transatlantic contexts. *Minerva* 52: 119–140.

Blanken, I.J. 2002. *Een industriële wereldfederatie: Geschiedenis van Koninklijke Philips Electronics N.V., vol. V.* Zaltbommel: Europese Bibliotheek.

Boersma, Kees. 2002. *Inventing structures for industrial research: A history of the Philips NatLab 1914–1946.* Amsterdam: Aksant.

Casimir, H.B.G. 1958. *Voorzieningen ten behoeve van de research binnen de faculteiten der wis- en natuurkunde der Nederlandse universiteiten.* The Hague: Staatsdrukkerij- en Uitgeverijbedrijf.

Collins, H., R. Evans, and M. Gorman. 2007. Trading zones and interactional expertise. *Studies in History and Philosophy of Science* 38: 657–666.

Dalum, Bent, Björn Johnson, and Bengt-Åke Lundvall. 1992. Public policy in the learning society. In *National systems of innovation: Towards a theory of innovation and interactive learning*, ed. Bengt-Åke Lundvall. London/New York: Pinter.

de Boer, Harry, and Jeroen Huisman. 1999. The new public management in Dutch universities. In *Towards a new model of governance for universities?* ed. D. Braun and F.X. Merrien, 100–118. London: Jessica Kingsley Publishers.

de Jager, Kees. 2009. ANS, de eerste Nederlandse satelliet. *Zenit*, p. 465.

de Kort, Niek. 2003. *Ruimteonderzoek: de horizon voorbij.* Amsterdam: Natuur & Techniek.

de Vries, Marc. 2005. *80 years of research at the Philips Natuurkundig Laboratorium 1914–1994.* Amsterdam: Pallas Publications.

Deterink, A.A.M., et al. 1997. *Onderzoek naar de oorzaak van het faillissement van Fokker.* Deventer: Kluwer.

Dierikx, Marc. 2004. *Uit de lucht gegrepen: Fokker als Nederlandse droom, 1945–1996.* Amsterdam: Boom.

Edgerton, David. 2004. The 'linear model' did not exist: Reflections on the history and historiography of science and research in industry in the twentieth century. In *The science—industry nexus: History, policy, implications*, ed. Karl Grandin and Nina Wormbs. New York: Watson.

Edqvist, Olle. 2003. Layered science and science policies. *Minerva* 41: 207–221.

Elbers, Astrid. 2012. The establishment of the new field of radio astronomy in the post-war Netherlands: A search for allies and funding. *Centaurus* 54: 265–285.

Etzkowitz, Henry. 2003. Innovation in innovation: The Triple Helix of university-industry-government relations. *Social Science Information* 42: 293–337.

Etzkowitz, Henry, and Loet Leydesdorff. 2000. The dynamics of innovation: From national systems and 'Mode 2' to a Triple Helix of university-industry-government relations. *Research Policy* 29: 109–123.

Flipse, A. 2012. 'Geen weelde, maar een offer'. De band tussen Vrije Universiteit en achterban, 1880–1950. In *Universiteit, Publiek en Politiek*, ed. L.J. Dorsman and P.J. Knegtmans, 67–82. Hilversum: Verloren.

Galison, Peter, and Bruce Hevly (eds.). 1992. *Big science: The growth of large-scale research*. Stanford: Stanford University Press.

Gibbons, Michael, Camille Limoges, Helga Nowotny, Simon Schwartzman, Peter Scott, and Martin Trow. 1994. *The new production of knowledge: The dynamics of science and research in contemporary societies*. London: Sage.

GROC. *Papers of the Geophysics and Space Research Committee*. Amsterdam: Royal Netherlands Academy of Sciences.

Heijmans, H.G. 1994. *Wetenschap tussen universiteit en industrie: de experimentele natuurkunde in Utrecht onder W.H. Julius en L.S. Ornstein 1896–1940*. Rotterdam: Erasmus Publishing.

Hoeneveld, F., and J. van Dongen. 2013. Out of a clear blue sky? FOM, the bomb and the boost in Dutch physics funding after World War II. *Centaurus* 55: 264–293.

Hollestelle, Marijn. 2011. *Paul Ehrenfest: Worstelingen met de Moderne Wetenschap, 1912–1933*. Leiden: Leiden University Press.

Johnson, Stephen. 2002. *The secret of Apollo: Systems management in the American and European space programs*. Baltimore: Johns Hopkins University Press.

Kersten, Albert. 1996. *Een organisatie van en voor onderzoekers: ZWO 1947–1988*. Assen: Van Gorcum.

Krige, John. 2006a. *American hegemony and the postwar reconstruction of science in Europe*. Cambridge, MA: MIT Press.

Krige, John. 2006b. Technology, foreign policy, and international cooperation in space. In *Critical issues in the history of spaceflight*, ed. Steven J. Dick and Roger D. Launius, 239–260. Washington, DC: NASA.

Krige, John. 2010. Building the arsenal of knowledge. *Centaurus* 52(4): 280–296.

Krige, John, and Arturo Russo. 2000. *A history of the European Space Agency 1958–1987, vol. I: The story of ESRO and ELDO 1958–1973*. Noordwijk: ESA.

Langford, Cooper H., and Martha Whitney Langford. 2000. The evolution of rules for access to megascience research environments viewed from Canadian experience. *Research Policy* 29: 169–179.

Lundin, Per, Niklas Stenlås, and Johan Gribbe. 2010. *Science for welfare and warfare: Technology and state initiative in Cold War Sweden*. Sagamore Beach: Science History Publications.

Muller, Ed (ed.). 1997. *50 jaar Nederlands Instituut voor Vliegtuigontwikkeling en Ruimtevaart*. Katwijk aan Zee: Satellite Services.

NA. National Archives, The Hague.

Nowotny, Helga, Peter Scott, and Michael Gibbons. 2003. Introduction: 'Mode 2' revisited: The new production of knowledge. *Minerva* 41: 179–194.

PCA. Philips Company Archives, Eindhoven.

Pestre, Dominique. 2003. Regimes of knowledge production in society: Towards a more political and social reading. *Minerva* 41: 245–261.

Scranton, Philip. 2006. Technology, science and American innovation. *Business History* 48: 311–331.

Smith, Robert W. 1989. *The space telescope: A study of NASA, science, technology, and politics.* Cambridge/New York: Cambridge University Press.

Spaa, J.H. 1975. Enige konklusies uit het ANS-projekt vanuit het Philips standpunt. Philips Company Archives, file 821:921.94 no. 4.

Sullivan, Woodruff T. 2000. Kapteyns influence on the style and content of twentieth century Dutch astronomy. In *The legacy of J.C. Kapteyn: Studies on Kapteyn and the development of modern astronomy*, ed. P.C. van der Kruit and K. van Berkel, 229–264. Dordrecht: Kluwer.

Sullivan, Woodruff T. 2009. *Cosmic noise: A history of early radio astronomy.* Cambridge: Cambridge University Press.

Tucker, Wallace, and Karen Tucker. 1986. *The cosmic inquirers: Modern telescopes and their makers.* Cambridge, MA: Harvard University Press.

van Berkel, Klaas. 2011. *De stem van de wetenschap: geschiedenis van de Koninklijke Nederlandse Akademie van Wetenschappen*, vol. 2. Amsterdam: Uitgeverij Bert Bakker.

van de Hulst, Henk. 1992. Seizing opportunities: Some comments on the Dutch national space science programme of the sixties and seventies. In *Science beyond the atmosphere: The history of space research in Europe*, ed. Arturo Russo, 125–138. Noordwijk: ESA.

van Helvoort, Ton. 2005. *De KNAW tussen wetenschap en politiek: de positive van de scheikunde in de Akademie in naoorlogs Nederland.* Amsterdam: KNAW.

van Otterloo, P. 1973. Management aspecten van het ANS project. Philips Company Archives, file 821:921.94 no. 4.

Waldrop, Mitchell. 1983. Infrared astronomy satellite. *Science* 220(4604): 1365–1368.

"We Will Learn More About the Earth by Leaving It than by Remaining on It." NASA and the Forming of an Earth Science Discipline in the 1960s

Roger D. Launius

8.1   Introduction

NASA has always been viewed by those observing it externally as the "space agency," and its leaders have long viewed it that way as well. Furthermore, agency personnel have historically defined NASA's human spaceflight efforts as its primary mission, with other activities as of lesser importance. I have described this as imprisonment in a prestige trap that constricts the agency, its leadership, and its range of options in charting a future in space.[1] Historian Paul Forman went further in characterizing scientists caught up in such government-supported endeavors as essentially gadgeteers not only seeking the use of instruments that the government was willing to develop for them but also channeling the scientific questions and investigations along lines where those priorities and capabilities

R.D. Launius (✉)
National Air and Space Museum, Smithsonian Institution, Washington, DC, USA

© The Editor(s) (if applicable) and The Author(s) 2016   211
T. Heinze, R. Münch (eds.), *Innovation in Science and Organizational Renewal*, DOI 10.1057/978-1-137-59420-4_8

could be maximized. In the process, equally valid questions that did not rise to this level of support were largely ignored.[2]

This perception has led such historians as Kim McQuaid to question NASA's commitment as an institution to other space activities, activities which in their view were more useful and therefore deserving of a higher priority.[3] While there is some basis for concluding that early on NASA missed an opportunity to dominate a very public effort to understand the Earth as a planet, in reality NASA officials pursued very important Earth science objectives in the 1960s that aided significantly in the rise of Earth system science in the decades that followed.[4] Indeed, at a fundamental level NASA—along with the National Oceanic and Atmospheric Administration (NOAA)—became the critical component in the 1960s of the origins of a new scientific discipline that emerged in the USA, Earth system science.

During that decade, NASA developed the critical technology—Earth-orbiting satellites capable of taking scientific readings on a global scale—that made possible the convergence of many different scientific disciplines into a single system of investigation. These remote-sensing satellites allowed study of Earth on a planetary scale for the first time, even though that effort was in its infancy and many of these satellites were experimental in nature. It represented the rise of interdisciplinarity in the various sciences focusing on understanding the Earth. As such, it incorporated understandings of how the atmosphere, ocean, land, and biospheric components of the Earth interacted as an integrated system. This resulted from studies of the interaction between the physical climate system and biogeochemical cycles. Very early the role of humans in this process emerged as NASA pursued research with its Landsat satellites to demonstrate changes in land use and ground cover. Only through the analysis of data obtained through both in situ observations and from remotely sensed observations, as well as the development of sophisticated ocean–atmosphere–land models, did this become possible. Not until the space age did a fundamental ingredient of this process emerge in the use of satellites. Earth system science, therefore, by the 1970s offered a foundation for understanding and forecasting changes in the global environment and regional implications.[5]

This approach, embracing as it did chemistry, physics, biology, mathematics, and other sciences, transcended disciplinary boundaries to treat the Earth as an integrated entity. NASA led the effort to bridge the divide between these disciplines and the increasingly interpretive and integrative endeavor for knowledge. By the mid-1980s, NASA's role in cataloging the elements of the Earth system, their connections, requirements, and changes had been fully realized.[6] By then, according to scientists Samuel N. Goward and Darrel L. Williams, it "was evident that satellite remote sensing could provide the type of globally consistent, spatially disaggregated, and temporally repetitive measurements of land conditions needed to describe the Earth's terrestrial systems. … These physical and biological processes are primary descriptors of how land conditions modulate the Earth system. Once it was understood that space-based, Earth imaging could and did provide such information about land patterns and dynamics, the possibility of developing fully integrated land-ocean-atmosphere monitoring and modeling capabilities was realized."[7]

Beforehand, of course, aerial photography had been employed for all manner of scientific efforts ranging from geology to climate monitoring to oceanography as part of interwar era programs. Likewise, overhead observation had been used since before World War I, and had become a standard by the Cold War; at the same time weather and climate scientists had used a broad array of platforms—ranging from balloons to aircraft—to gather readings about the atmosphere, fronts, and patterns from the dawn of the twentieth century.[8] It took some investment in technology and cultural changes to accept space observation into these scientific disciplines, and in 1962, NASA sponsored its first conference discussing the possibilities of space-based observations, less than four years after the birth of the agency. The first programs in these arenas proved successful beginning in the 1960s, all of them coming with considerable investment in the development of new technology and in the persuasion of potential users that they could benefit from space-based scientific data. A small group of NASA officials working on space science and applications programs led the effort to develop these resources.[9]

In addition, NASA pursued a large-scale effort to lay the groundwork in Earth system science at its Goddard Space Flight Center's Division of Aeronomy and Meteorology under William Stroud. The agency's managers nurtured scientific activities in this realm, and worked effectively to create networks of researchers who had strong interest in using remote-sensing technology to observe and measure aspects of the Earth's climate from space. Without question, data from NASA technology, satellites, institutes, scientists, and organizational initiatives were essential in creating the global picture of the Earth as a system that emerged later.[10]

There was, and remains, a tension between NASA's human spaceflight enthusiasts and their relative disinterest in studies of the Earth, and the researchers interested in science and answers to scientific questions. In NASA those people tended to be self-sorted into the agency's two big efforts, the human program and the science program. They competed with each other for resources and a small part of the science program emphasized studies of the Earth. Interestingly, NASA pursued this Earth science agenda, albeit initially at a modest level, in spite of official circumscriptions on its mission. The National Aeronautics and Space Act of 1958 did not assign a broad mandate for studies of the Earth, focusing instead on "The expansion of human knowledge of phenomena in the atmosphere and space."[11] The explicit mention of the atmosphere provided a crack in the door allowing NASA's Office of Space Sciences and Applications to pursue Earth observations from space as well as to support suborbital stratospheric research. It also enabled NASA to work with other organizations, in this case the Weather Bureau, to develop technology programs that supported its mission within the "applications" portion of NASA's mission. These twin prongs of NASA's efforts led eventually to the agency's leadership of a broad-based Earth system science effort by the 1980s.[12]

The editors of this volume argue that one institutional condition for progress and renewal in science is "organization of interdisciplinary research" (see editor's introduction). At a fundamental level, NASA leaders during its early years encouraged precisely the kind of collaboration between scientists from many different disciplines focused on the Earth—geology, atmospherics and climatology, oceanography, biology, chemistry, and physics—that transcended disciplinary boundaries using space technology to treat the Earth as an integrated system. The revolution in understanding coming through this process was profound, and without the leadership of key NASA individuals and organizations it is problematic if such an alteration could have been affected on the schedule in which it took place. And this took place in an institution not predisposed to focus on that activity as it undertook the race to the Moon. It highlights that leadership at NASA, that helped to form a broad-based, multidisciplinary community of scientists, oriented toward understanding planet Earth in much the same way that it sought to understand other planets in the solar system and what this portended for the future of this scientific activity. Indeed, from limited cooperative efforts in the 1960s and 1970s overseen by NASA, emerged the broadly interdisciplinary efforts to understand the interactions determining the past, present, and future of Earth science of the last quarter century.

During its formative era, even as it undertook the Moon landing program that became its signature accomplishment, NASA helped to shape the structure of the Earth sciences. In essence, NASA's role in this arena—in contrast to the National Science Foundation, the Department of Defense (DoD), and the NOAA—aimed at fostering collaborative, multidisciplinary investigations at the macro level afforded by the capabilities of space technology. The process was never easy, as NASA and other entities jockeyed for position/influence/suzerainty over the course of the field. Often sheer power—especially in the form of money for projects—dominated the course of these relationships. Often NASA, as a well-funded US governmental agency, was able to gain the upper hand for its priorities. Through the process the longstanding direction of Earth science was charted, for good or ill.

8.2   NASA, THE NATIONAL WEATHER BUREAU, AND THE DEVELOPMENT OF WEATHER SATELLITES

In a recent opinion piece in the *New York Times*, Heidi Cullen commented on the critical importance of weather satellites overhead and how they have enhanced the public good for more than 50 years:

> We have made tremendous progress in the accuracy of our hurricane forecasting (and overall weather forecasting) since then, much of it a result of government-owned satellites that were first launched in the 1960s and now provide about 90 percent of the data used by the National Weather Service in its forecasting models. Satellite and radar data and the powerful computers that crunch this information are the foundation of the weather information and images we get. Thanks to these instruments, for instance, the five-day hurricane track forecast we get today is more accurate than the three-day forecast from just 10 years ago.[13]

This did not happen by magic; instead it required leaders from several communities—users, scientists, and engineers, as well as others—to establish what has now become the norm for weather forecasting and climate data collection from remote-sensing satellites.

It was obvious even before the Space Age that satellites would be useful for meteorology. The classic 1946 study, Project RAND's "Preliminary Design of an Experimental World-Circling Spaceship," argued that one of the two key uses of a satellite would be for "observation of weather conditions," noting that tracking cloud patterns "should be of extreme value in connection with short-range weather forecasting, and tabulation of such data over a period of time might prove extremely valuable to long-range weather forecasting."[14] These analyses, coupled with meteorological data obtained from suborbital rockets launched in the American Southwest beginning in 1947, led to a consensus from scientists and engineers alike that weather satellites possessed great promise for the future.[15]

Harry Wexler, the Weather Bureau's Chief of Scientific Services, recognized there were important uses for weather satellites in forecasting patterns, even though he did not envision a satellite's potential for what became routine observations of pressure, temperature, wind velocity, and humidity. He wrote in 1954 that a satellite could provide a valuable "bird's eye" view of weather patterns, but would be less useful in obtaining the "three-dimensional data" meteorologists needed, relegating the technology to use for tracking violent weather patterns and "storm patrol."[16] It was this limitation of satellite observation, furthermore, that prompted efforts to develop ever more sophisticated weather satellites in the 1960s and 1970s.

Even as this was underway, in July 1958 President Dwight D. Eisenhower decided that all space programs that were not clearly military should be transferred to the new civilian space agency and he assigned the TIROS weather satellite program to NASA upon its activation on October 1, 1958. At near the same time, the Eisenhower administration designated the Weather Bureau as "their meteorological agent in providing the meteorological instrumentation, data reduction and analysis of observations taken by satellites after the International Geophysical Year (IGY) Series is finished."[17] This was part of a larger effort to ensure that space be viewed as a non-threatening environment for the Soviet Union; admittedly this characterization was a bit of a ruse but it served the need of appearing to emphasize peaceful purposes for space activities. The transfer posed challenges since the program was so far along in its development, but a number of scientists and engineers agreed to move from the DoD to NASA along with the project, and NASA negotiated an effective agreement with the Weather Bureau to provide meteorological research support. Thereafter, the DoD remained involved but not in the forefront of research, development, and operations.[18]

NASA launched TIROS 1 on April 1, 1960, and it proved successful from the outset, despite technical problems and difficulties in working across several federal agencies. "Two television cameras looking down from an altitude of about 450 miles made initial pictures of the earth's cloud patterns during the satellite's second orbital trip," reported the *New York Times* just after the launch. Unveiled by NASA, as the federal agency responsible for the TIROS program, the representatives of the Weather Bureau and the Eisenhower administration gushed about the prospects for future observation of weather patterns and better forecasting an operational weather satellite system would provide.[19] NASA administrator T. Keith Glennan wrote in his diary about meeting with Wexler and others after the first TIROS launch and expressed concern about how best to characterize the mission. "The Weather Bureau people are apt to be a little enthusiastic—more than enthusiastic—about the prospects that are available to us with TIROS," he commented. "They all agreed, finally, to play down these stories and to be as factual as possible in their discussion of TIROS."[20] Despite the soft pedaling, the satellite provided valuable images of weather fronts, storms, and other atmospheric occurrences. It led directly to a long series of weather satellites that quickly became standard weather forecasting tools in the USA and throughout the world. TIROS helped meteorologists forecast patterns and study weather and climate. Placed in a polar orbit, TIROS proved Wexler wrong in relegating weather satellites to the role of only "storm chasers" as it fundamentally altered both scientific and practical applications.

With the success of TIROS, NASA and the Weather Bureau embarked on a succession of experimental weather satellites, some named TIROS but also a second-generation satellite called Nimbus. More complex than TIROS, Nimbus carried advanced TV cloud-mapping cameras and an infrared radiometer that allowed pictures at night for the first time. Seven Nimbus satellites were placed in orbit between 1964 and 1978, creating the capability to observe the planet 24 hours per day. Turning weather satellites from an experimental program to an operational system proved daunting. To accomplish this, NASA and Weather Bureau scientists organized an interagency Panel on Operational Meteorological Satellites in October 1960. Developers, scientists, other users, and various federal agencies aired disagreements over the future of the program in this setting; the meetings were often contentious. The Weather Bureau sought complete authority over the planned operational system, including launching, data retrieval, and final decisions on the design of new operational satellites. The key provision for the Weather Bureau gave it "program responsibility for the operational meteorological satellite observing and data processing system. This would include equipment procurement, launching, data retrieval and processing, and dissemination to users. An organization to perform all activities related to the operational meteorological satellite observing system would be established as a self-contained entity reporting to the Chief of the Weather Bureau."[21] NASA officials could never agree to such a situation and pushed back. Keith Glennan at NASA remarked in his diary that it was an off-putting task to deal with "the problems that face us in developing a meteorological satellite system that will be operational within the next four or five years. It is so obvious that the Weather Bureau is poorly prepared to take on the research necessary to deal with this very difficult problem, one wants to step in and help. Unfortunately, it doesn't appear as though we'll be able to do very much because we don't have too much of the necessary experience."[22]

The Weather Bureau, NASA, and the DoD agreed to a compromise in April 1961 that endorsed a national operational meteorological satellite system based on a second-generation satellite already under NASA development which would be managed by the Weather Bureau. In this plan, NASA retained control of launch services and ground support for the system as well as the R&D on the spacecraft. The Weather Bureau had responsibility for operations, data storage and analysis, and communication of all results.[23]

The ESSA[24] 1 through 9 satellites provided some upgrade to what had gone before. Additionally, meteorological satellites that were part of NASA's Applications Technology Satellite (ATS) project to orbit experimental geosynchronous satellites proved valuable. In December 1966 and November 1967, ATS 1 and 3 explored the possibility of observing weather with line scan imagers; the resulting continuous coverage proved valuable for short-lived cloud patterns correlated to tornadoes. Continuous coverage from geosynchronous orbit made it possible to observe the motion of clouds and deduce wind speed at the level of the clouds. Three other satellites launched in the 1960s, ATS 2, 4, and 5, also carried out meteorological experiments. In addition, NASA pursued for the Weather Bureau a series of weather satellites for its operational system. In 1966, for example, Robert M. White, Administrator of the Environmental Science Services Administration, told NASA's Homer Newell that his agency was looking forward to working with NASA to "satisfy environmental data requirements in other areas than meteorology. … We have been most pleased with the joint effort this past year in resolving problems and in allocating available resources to meet operational and R&D meteorological satellite needs." He added, "We are looking forward to the continuation of this excellent cooperation."[25] The result was by the end of the 1960s, NASA had worked with the Weather Bureau to develop sophisticated technologies aimed at completing the desired operational system, and while the relationships were always rocky both sides made it work to accomplish useful ends.

From this story of the development and deployment of weather satellites in the 1960s, three primary conclusions emerge. First, neither NASA nor the Weather Bureau had experience in cooperating with other organizations in the accomplishment of the mission to forecast the weather and gain knowledge of the climate. They had to learn those skills, and to a greater or lesser degree they did.[26] Both groups had to build relations that were effective, and regardless of the roughness of the road at times they generally accomplished this feat over the course of the decade. It was very much, however, a process of two steps forward and one step backwards. In the process, NASA gained an entrée into a major scientific enterprise that has expanded in the decades since that early era of the Space Age and climatologists and meteorologists gained a valuable new source of data.

Second, the story of weather satellites is one in which the longstanding tensions between scientists and engineers played out. Getting all elements to work together challenged every program, regardless of whether or not they were civil service, industry, or university personnel, but the divergent organizational and professional cultures of the two groups led to constant difficulties. They differed over priorities and competed for resources. The two groups contended with each other over a great variety of issues associated with the TIROS and Nimbus programs. For instance, the scientists disliked having to configure payloads so that they could meet time, money, or launch vehicle constraints. The engineers, likewise, resented changes to scientific packages added after project definition because these threw their hardware efforts out of kilter. Both had valid complaints and had to maintain an uneasy cooperation to accomplish the tasks at hand.[27]

Third, these weather satellite efforts demonstrated the significance of this capability to greater understanding of the Earth. This was obvious from the very first launch of TIROS 1 in 1960. While satellite networks would never supplant conventional ground observation, they proved a valuable means of expanding conventional approaches to weather data collection.[28] The result was an awakening at NASA during the early- to mid-1960s of the significance of this arena of space science and applications. While it never dominated the agency, and there was resistance to it in some quarters, the seeds of the Earth system science discipline were planted during this era.

8.3   REPLICATING PLANETARY SCIENCE
FOR PLANET EARTH

In addition to weather satellite development, deployment, and operations in the 1960s, NASA scientists under the leadership of Homer Newell, the agency's Associate Administrator for Space Science, began a serious effort toward creating Earth system science through the realization that Earth was a planet and could be studied in the same way scientists were doing elsewhere in the solar system. The IGY of 1957–1958 had pioneered this development, and NASA furthered it in the early part of the 1960s.[29]

Using satellites, scientists were able in the 1960s to undertake path-breaking geodetic research. They measured the Earth as never had been possible before. By 1970 a worldwide geodetic net had been established, allowing common reference points to be established anywhere on the globe with an accuracy of 15 meters. An important outgrowth of this satellite research, although other scientific approaches also contributed, was the theory of plate tectonics advanced in the latter 1960s to explain the dynamics of the Earth's outer shell. The theory posited that the Earth's surface, the lithosphere, consists of about a dozen large plates and several smaller ones that moved relative to each other and interacted at their boundaries. This theory went far toward explaining seismic and volcanic activity as well the origins and evolution of mountains and other geographical features.[30]

Atmospheric, ionospheric, and geophysical science also benefitted greatly from the opportunity to study the Earth from satellites. Building steadily on the research base of earlier years, as Homer Newell wrote, by the end of the decade of the 1960s "all known major problems of the high atmosphere and ionosphere had a satisfactory explanation based on sound observational data."[31]

The genesis of planetary Earth studies came via a means not easily envisioned at NASA—from its planetary scientists at universities around the country. Early on NASA recruited a small cadre of scientists interested in planetary climatology, geology, geodesy, biology, chemistry, and magnetospherics to focus on studies of Venus and Mars but everyone realized the same instruments could be used to obtain a planetary perspective on Earth. As scientist Harrison Brown wrote in a National Research Council (NRC) study in 1961:

> Until recently man has been confined to the Earth's surface, with the result that the types of observation which he has been able to make have been

severely limited. The airplane liberated him to some extent and made possible a variety of photographic and meteorological measurements. High altitude rockets increased further his capabilities for measurement. But the development of satellites and space probes without question is adding a new dimension to his capabilities. It seems likely that in the years ahead we will learn more about the Earth by leaving it than by remaining on it.[32]

It was clear, as stated in a 1962 NRC report, that the entrée of scientists into Earth observation came because of the desire to focus on Venus and Mars. "Much of our knowledge of the planets has been and will continue to be based on lessons learned from studying our own planet." The report concluded, "With this in mind, it is clear that no opportunity should be lost to test out planetary probe experiments from rockets and Earth satellites. In addition to serving as 'field tests' for new equipment and techniques, these tests can be valuable scientific experiments in their own right, and in all likelihood will give vital information about our own planet."[33]

In terms of developing an Earth science community, NASA also incorporated the scientific community into mission planning, instrumentation building and use, and data collection and analysis.[34] Many individual scientists migrated from planetary to Earth science over time as the 1960s progressed, helping to create Earth science as a cohesive entity. A few examples demonstrate this transition. Alan H. Barrett was a physicist with a B.S. from Purdue University and an M.S. and Ph.D. from Columbia University who engaged in a broad set of studies first as a fellow at the Naval Research Laboratory and later as a research associate and instructor in astronomy at the University of Michigan before moving to the Massachusetts Institute of Technology (MIT) in 1961. As an astronomer Barrett was brought by NASA into the research team working on Mariners 1 and 2; while Mariner 1 failed, Mariner 2 became the first mission in 1962 to fly-by the planet Venus. Barrett served as principal investigator for the microwave radiometer on these spacecraft that found surface temperatures there far too high to support life as known on Earth.[35] He continued with astronomical research but also got involved in Earth science:

In 1963 Dr. Barrett, together with associates at M.I.T, reached a milestone in radio astronomy by detecting and measuring the presence of hydroxyl in interstellar space, using an 84-foot-diameter instrument at the university's Lincoln Laboratory on Millstone Hill. It was the first time a molecule had been found in the Milky Way and the discovery opened the way for development of a new field of research, the study of molecules in the far regions of the universe.[36]

Later in his career, Barrett transitioned to Earth science studies, using the very same approach to radio astronomy used in his earlier work "to study characteristics of the Earth's atmosphere. That work prepared the way for the Nimbus series of meteorological satellites."[37] He also trained a generation of students in planetary science, such as the still active David H. Staelin, who also concentrated in Earth science studies.[38]

A second example was Conway B. Leovy, who received his Ph.D. in meteorology from MIT in 1963, and pursued atmospheric studies and climatology first at the RAND Corp. and after 1968 on the faculty of the University of Washington. Throughout this period, Leovy participated in imaging experiments on NASA's Mariner 6 and 7 fly-by missions to Mars, which were followed, in the 1970s, with participation in the imaging experiment of the Mariner 9 Mars orbiter and the meteorology experiment on NASA's Viking landers. This set him on a path of Mars research that he was involved in throughout his career, but he also soon broadened this to the study of atmospheres for all of the planets of the solar system, including Earth. "During these early years," wrote a colleague at the time of his death in 2011, "Conway furthered our understanding in different branches of atmospheric science: the chemistry of atmospheric ozone, the behavior of convection near the surface of the Earth, and the energy budget and motion of the air in Earth's mesosphere (about 55–85 km altitude)." Leovy published a large number of papers about Mars, about the Earth, and about Jupiter, Venus, and Titan. Through this process he dreamed that "a general theory of atmospheric dynamics might be developed that could explain the observed winds, temperatures and energy distributions in each planetary atmosphere. But when the data came back from spacecraft in the 1970s and 1980s, he came to realize that planetary atmospheres are so strange and diverse that a group of general principles rather than a general theory was the more practical pursuit." Largely identified with planetary atmospheres research—training no fewer than 23 Ph.D.'s in this field—he also was heavily involved in Earth science with his first paper on the subject published in 1964 and the last one in 2000.[39]

Third, Hugh R. Anderson—a son of a professor at the University of Iowa—earned a B.A. and M.S. in physics in Iowa City and then went on to the California Institute of Technology where he completed his Ph.D. in physics in 1961. While on active duty with the US Air Force, Anderson was assigned to the Jet Propulsion Laboratory, and when he finished his military commitment he stayed at JPL as a scientist in the Experimental Space Sciences Section. He worked on both Mariners 2 and

4, mostly contributing to particles and fields research in interplanetary space. Indeed, that subject was dominant throughout his career, but he also studied the relationship of the Earth to cosmic energy and made critical contributions to solar energy and its relationship to the planet's "polar caps."[40] Furthermore, in a paper coauthored with Conway Snyder, Marcia Neugebauer, and Edward J. Smith, they observed: "It must be remembered that one of the most effective ways to study interplanetary entities has been, and in the Space Age still is, through the observation of their effects upon the Earth."[41]

Finally, there is the fascinating story of S.I. Rasool, born in India in 1930 and earning his Ph.D. in Atmospheric Sciences at the University of Paris in 1956. He worked in both the USA and Europe during his career. He served on the science team for the Mariner Mars effort in the latter 1960s, and transitioned to Earth science full time thereafter.[42] He has been involved in climate change research since the 1970s; his work has included path-breaking papers exploring the Earth as a system.[43] He argued repeatedly over the years:

> There is now compelling evidence that man's activities are changing both the composition of the atmospheric and the global landscape quite drastically. The consequences of these changes on the global climate of the 21st century is currently a hotly debated subject. Global models of a coupled Earth-ocean-atmosphere system are still very primitive and progress in this area appears largely data limited, specially over the global biosphere. A concerted effort on monitoring biospheric functions on scales from pixels to global and days to decades needs to be coordinated on an international scale in order to address the questions related to global change.

He emphasized the need for obtaining coordinated, long-term data on the changing nature of the planet's climate to understand the full nature of what has been taking place.[44] Rasool co-wrote a paper in *Science* in 1971 that applied climate models developed for understanding Venus to Earth and came up with a surprising conclusion:

> Effects on the global temperature of large increases in carbon dioxide and aerosol densities in the atmosphere of Earth have been computed. It is found that, although the addition of carbon dioxide in the atmosphere does increase the surface temperature, the rate of temperature increase diminishes with increasing carbon dioxide in the atmosphere. For aerosols, however, the net effect of increase in density is to reduce the surface temperature of

Earth. Because of the exponential dependence of the backscattering, the rate of temperature decrease is augmented with increasing aerosol content. An increase by only a factor of 4 in global aerosol background concentration may be sufficient to reduce the surface temperature by as much as 3.5 K. If sustained over a period of several years, such a temperature decrease over the whole globe is believed to be sufficient to trigger an ice age.[45]

Almost immediately, *Washington Post* reporter Victor Cohn published a story about this scientific paper, emphasizing its most dramatic aspects—the prospect of radical temperature drop and potential ice age. Among other quotes in the news report—commenting on the possibility that if human-made aerosols increased by a factor of four, they might cause massive global cooling—was an aside by Rasool that the Earth "could be as little as 50 or 60 years away from a disastrous new ice age."[46] Rediscovered in the 1990s, this *Science* article—and especially its popularization—became the starting point for the feature film, *The Day After Tomorrow*, a film that the *Guardian* characterized as "a great movie and lousy science."[47] It has also been used by global warming skeptics to counter dire predictions of the future.

To these individuals, we could add many others, some of them luminaries such as Carl Sagan, who contributed in both planetary and Earth science over their careers.[48] One of these individuals, Robert P. Sharp, a Caltech geologist "applied the lessons offered by a close study of Earth to the challenge of understanding other planets." He served as a team scientist on the Mariner 4 (1965), 6, 7 (1969), and 9 (1971) flights to Mars, applying his knowledge of terrestrial landforms and processes to the study of the surface of Mars. At a 1966 conference, Sharp praised scientists for becoming more involved in the planetary program, but urged them, too, to "look downward into our own planet ... Our understanding of these distant bodies will depend to a good degree upon how well we understand our own plain Earth."[49]

At sum, during the 1960s these activities sponsored by NASA to explore the other planets of the solar system had also built up a cadre of researchers also interested in deploying satellites to investigate the Earth as a planet. This became the starting point for the development of a new cross-disciplinary scientific effort that has come to be called since the 1980s Earth system science. Far from ignoring "earthly environmentalism," as some have argued, NASA fostered this growing field of study through its unique vantage point of scientific investigation from orbit. It was never the agency's primary focus, of course, but it took place nonetheless. It is, in essence, a serendipity of the investment in space science and technology undertaken by the USA during the space race to the Moon with the Soviet Union.

8.4   The Organizational Imperative

When NASA was established on October 1, 1958, Administrator T. Keith Glennan had to build from scratch a space science program, hiring administrators and scientists to work at the new agency as well as engaging those at universities to participate in the program. Glennan emphasized that NASA should be responsible for space science and created the Office of Space Flight Programs at NASA Headquarters under the leadership of Abe Silverstein, a propulsion engineer from the NACA Lewis Research Laboratory. Silverstein made the key decision of bringing Homer Newell over to NASA from the Naval Research Laboratory, along with 50 other scientists, to shape the new space science program for the agency.[50] NASA would ask educational and research institutions, industry, and federal laboratories to participate in the program.[51] By 1963, this structure had morphed into the Office of Space Science and Applications with Homer Newell as Associate Administrator reporting directly to the NASA administrator.[52] This was the structure created and sustained for more than decade in which the rise of Earth sciences grew into a form that it might be able to accomplish the beginnings of this mission. Later, separate organizations would be established to oversee technology development and operational activities; and much later in the latter 1980s a separate entity reporting to the administrator would be established to oversee Earth science activities.

All the while, NASA's dominant organizational culture and mission orientation inhibited efforts in its first decade to focus much attention on Earth science although there were firm pushes in that direction.[53] Over time, the Earth science aspects of the NASA mission have emerged as a critical component of the agency's activities. By the early 1970s, the Earth sciences enjoyed heightened visibility within NASA, as the Landsat Earth monitoring program became operational.[54]

Although not initially viewed as a science program, but rather a technology demonstrator, Landsat 1's launch on July 23, 1972, changed the way in which many people viewed the planet. It provided data on vegetation, insect infestations, crop growth, and associated land use. Two more Landsat vehicles were launched in January 1975 and March 1978, performed their missions effectively, and exited service in the 1980s.

Success in a "second generation" Landsat spacecraft followed in the 1980s with greater capabilities to produce more detailed land-use data. The system enhanced the ability to develop a worldwide crop forecasting system. Moreover, Landsat imagery has been used to devise a strategy for deploying equipment to contain oil spills, to aid navigation, to monitor pollution, to assist in water management, to site new power plants and pipelines, and to aid in agricultural development.[55]

By the 1970s, such programs as Landsat, and the Large Area Crop Inventory Experiment that was an Earth observation project using Landsat satellites to gather data, were becoming indispensable. So too, was a relatively small project to study stratospheric ozone depletion within the NASA science organization. In part, this resulted from the Space Shuttle's own potential to deplete ozone, but this initiative became politically salient very rapidly as the first of the American "ozone wars" broke out around chlorofluorocarbons.[56] James C. Fletcher, outgoing NASA administrator in 1977, remarked that these efforts represented the "'wave of the future' as far as NASA's public image is concerned. It is the most popular program (other than aeronautics) in the Congress and as you begin to visit with community leaders, you will understand it is clearly the most popular program with them as well."[57] These efforts in the 1970s rested firmly on the base established in earlier era. By the end of that decade, NASA had committed more funding to Earth science than any other federal organization, and its organization structure had evolved to oversee expansive scientific investigations across a broad spectrum of disciplines and technologies.[58]

Indeed, the results of Landsat proved integral to the official establishment of Earth systems science in the 1980s. As Goward and Williams concluded:

> The Landsat series of satellites constitute an explicit and integral component of the U.S. Global Change Research Program and, as described within this paper, helped to lead the scientific research community to develop and expand the concept of Earth Systems Science over the past two decades. The Landsat satellites have also provided data to a broad and diverse constituency of users who apply the data to a wide spectrum of tasks. This constituency encompasses the commercial, academic, government (federal, state, local), national security, and international communities. As of mid-1997, the Landsat satellites will have provided a continuous and consistent 25-year record of the Earth's continental surfaces that is unique and invaluable, and continuation of this database is critical to our global change research strategy.[59]

While Landsat was not the sole means whereby the rise of Earth system took place, its success was integral to it.

8.5   CONCLUSION

It was only with the broad developments over some 20 years beforehand that NASA was able to take the step it pursued in 1986 to establish a formal "Mission to Planet Earth" (MTPE) program. This came the aftermath of the *Challenger* accident in January 1986 when NASA commissioned astronaut Sally Ride to undertake a study of NASA programs and recommend an approach for future missions. *NASA Leadership and America's Future in Space: A Report to the Administrator* appeared in 1987. The so-called "Ride Report" proposed four main initiatives for study and evaluation:

1. MTPE
2. Exploration of the Solar System
3. Outpost on the Moon
4. Humans to Mars

The "Mission to Planet Earth" initiative called for the expansion of Earth science and the application of new technologies to understand the Earth as an integrated whole and the changes that may be taking place on it.[60]

While there had to be rescoping of the program over time, this report served as the catalyst for an investment of more than $7 billion to build and operate a series of orbital spacecraft, and to analyze data from them for environment purposes. The program's Earth Observing System satellites consisted of a range of remote-sensing satellites that collected data in a variety of ranges on air, land, and sea bodies on the planet. In 1991, NASA formally established MTPE as a comprehensive program for studying Earth from space. It emphasized the integration of data from various Earth-observing instruments and programs to gain a greater understanding of Earth's natural processes on a global scale. The perspective provided new levels of precision to the evaluation of pressure fronts and air masses that are so critical in weather forecasting. Likewise, meteorological research beyond weather forecasting took on new life as climatological research contributed significant insights into the understanding of Earth.[61]

By 2000, Earth system science had matured and a variety of Earth-observing spacecraft were enabling scientists to obtain sophisticated data about this planet's physical characteristics. Among others, these spacecraft included the Tropical Rainfall Measuring Mission, the Sea-viewing Wide Field-of-view Sensor mission, the QuikSCAT and TOPEX/Poseidon ocean studies missions, and the Active Cavity Radiometer Irradiance Monitor Satellite and Upper Atmosphere Research Satellite missions. Instruments from these satellites are measuring atmospheric chemistry, biomass burning, and land-surface changes ranging from Greenland to the tropical Pacific Ocean. Collectively, these spacecraft have revolutionized our understanding of the Earth. Collectively, they have shown changes in the atmosphere, land, and oceans, as well as their interactions with solar radiation and with one another.[62]

The foundation for the full-blown emergence of Earth system science was laid at NASA in the early 1960s and was nurtured and brought to fruition in the succeeding decades. The record is clear in terms of both climate and atmospheric science and planetary science turning toward issues of the Earth during those early years. Over time, this built into a fundamental structure of interdisciplinarity and institution building—what the editors of this volume call "investments in exploration." This chapter has focused on NASA's institutional issues associated with conducting Earth science in an organization that was predisposed not to be focused on that activity as it undertook the race to the Moon. It has highlighted some leadership at NASA that helped to form a broad-based, multidisciplinary community of scientists oriented toward understanding planet Earth in much the same way that it sought to understand other planets in the solar system and what this portended for the future of this scientific activity.

NOTES

1. Roger D. Launius, "Imprisoned in a Tesseract: NASA's Human Spaceflight Effort and the Prestige Trap," *Astropolitics: The International Journal of Space Politics and Policy* 10, no. 2 (2012).
2. Paul Forman, "Behind Quantum Electronics: National Security as Basis for Physical Research in the United States, 1940–1960," *Historical Studies in the Physical and Biological Sciences* 18, no. 1 (1987): 149–229.
3. Kim McQuaid, "Selling the Space Age: NASA and Earth's Environment, 1958–1990," *Environment and History* 12 (2006): 127–28.
4. See Erik M. Conway, *Atmospheric Sciences at NASA: A History* (Baltimore, MD: Johns Hopkins University Press, 2008), 64–93.
5. Veronica Boix Mansilla, "Learning to Synthesize: The Development of Interdisciplinary Understanding," in *The Oxford Handbook of Interdisciplinarity*, ed. Robert Frodeman et al. (New York: Oxford

University Press, 2010); Harry Collins, Robert Evans, and Mike Gorman, "Trading Zones and Interactional Expertise," *Studies in History and Philosophy of Science Part A* 38, no. 4 (2007): 657–66.

6. Naotatsu Shikazono, *Introduction to Earth and Planetary System Science: New View of Earth, Planets, and Humans* (New York: Springer, 2012), 8; Wolfgang Sachs, *Planet Dialectics: Explorations in Environment and Development* (London, UK: Zed Books, 1999), 101, 120–21, 125.

7. Samuel N. Goward and Darrel L. Williams, "Landsat and Earth Systems Science: Development of Terrestrial Monitoring," *Photogrammetric Engineering & Remote Sensing* 63 (July 1997): 887–88.

8. Tom D. Crouch, *The Eagle Aloft: Two Centuries of the Balloon in America* (Washington, DC: Smithsonian Institution Press, 1983); L.T.C. Rolt, *The Aeronauts: A History of Ballooning—1783–1903* (New York: Walker and Company, 1966); John H. Morrow Jr., *The Great War in the Air: Military Aviation from 1909–1921* (Washington, DC: Smithsonian Institution Press, 1993); Guilllaume de Syon, *Zeppelin! Germany and the Airship, 1900–1939* (Baltimore, MD: Johns Hopkins University Press, 2002); Charles C. Bates and John F. Fuller, *America's Weather Warriors, 1814–1985* (College Station: Texas A&M University Press, 1986).

9. Conway, *Atmospheric Science*; Helen Gavaghan, *Something New Under the Sun: Satellites and the Beginning of the Space Age* (New York: Copernicus Books, 1998); Janice Hill, *Weather from Above* (Washington, DC: Smithsonian Institution Press, 1991); Frederik Nebeker, *Calculating the Weather: Meteorology in the 20th Century* (San Diego, CA: Academic Press, 1995); P. Krishna Rao, *Evolution of the Weather Satellite Program in the U.S. Department of Commerce: A Brief Outline* (Washington, DC: NOAA, 2001); William K. Stevens, *The Change in the Weather: People, Weather, and the Science of Climate* (New York: Delacorte Press, 1999).

10. L. C. Nkemdirim, "The Global Atmospheric Research Program and the Geographer," *The Professional Geographer* 27 (1975): 227–30; R.J. Polavarapu and G.L. Austin, "A Review of the GARP Atlantic Tropical Experiment (GATE)," *Atmosphere-Ocean* 17, no.1 (1979): 2–13; Y.P. Borisenkov, "Global Atmospheric Research Program (GARP)," *Meteorology and Hydrology*, no. 12 (1974): 155–62; National Research Council, "National Academy of Sciences Annual Report, Fiscal Year 1968–69," 1969, 26–27.

11. "National Aeronautics and Space Act of 1958," Public Law#85-568, 72 Stat., 426, signed July 29, 1958, accessed 6/3/2012 11:18 AM, http://history.nasa.gov/spaceact.html.

12. Erik M. Conway, "Earth Science and Planetary Science: A Symbiotic Relationship?," in *NASA's First 50 Years: Historical Perspectives*, ed. Steven J. Dick (Washington, DC: NASA SP-2010-4070, 2010).

13. Heidi Cullen, "Clouded Forecast," *New York Times*, May 31, 2012.
14. Douglas Aircraft Company, Inc., "Preliminary Design of an Experimental World-Circling Spaceship," Report No. SM-11827, May 2, 1946, pp. 11, 13, copy available in NASA Historical Reference Collection, NASA History Office, NASA Headquarters, Washington, D.C.
15. Otto E. Berg, "High-Altitude Portrait of Storm Clouds," *Office of Naval Research Reviews*, September 1955, NASA Historical Reference Collection; Merton E. Davies and William R. Harris, *RAND's Role in the Evolution of Balloon and Satellite Observation Systems and Related U.S. Space Technology* (Santa Monica, CA: The RAND Corporation, 1988), 22.
16. Harry Wexler, "Observing the Weather from a Satellite Vehicle," *Journal of the Interplanetary Society* (September 1954): 269.
17. Harry Wexler, "Satellite Meteorology," September 10, 1958, attached to Rao, "Evolution of the Weather Satellite Program in the U.S. Department of Commerce," 20–25.
18. Hugh L. Dryden, for T. Keith Glennan, NASA, and Roy W. Johnson, Department of Defense, Agreement Between the Department of Defense and the National Aeronautics and Space Administration Regarding the TIROS Meteorological Satellite Project, April 13, 1959, NASA Historical Reference Collection; Chapman, "Case Study," 60–64.
19. Richard Witkins, "U.S. Orbits Weather Satellite; It Televises Earth and Clouds; New Era in Meteorology Seen," *New York Times*, April 2, 1960, 1.
20. J.D. Hunley, ed., *The Birth of NASA: The Diary of T. Keith Glennan* (Washington, DC: NASA SP-4105, 1993), 117.
21. US Department of Commerce, Weather Bureau, "National Plan for a Common System of Meteorological Satellites," Technical Planning Study No. 3, Preliminary Draft, October 1960, NASA Historical Reference Collection.
22. Hunley, *The Birth of NASA*, 243.
23. Ibid., 277; US National Coordinating Committee for Aviation Meteorology, Panel of Operational Meteorological Satellites, *Plan for a National Operational Meteorological Satellite System* (Washington, DC: US Government Printing Office, 1961).
24. The Weather Bureau had become part of a new organization, the Environmental Science Services Administration (ESSA), established on July 13, 1965, hence the name.
25. Robert M. White, Administrator, Environmental Science Services Administration, National Environmental Satellite Center, Department of Commerce, to Dr. Homer E. Newell, Associate Administrator for Space Science and Applications, NASA, August 15, 1966, NASA Historical Reference Collection; Smith, "The Meteorological Satellite," 45562; Hill, *Weather from Above*, 2326, 2935.

26. S. Fred Singer, Meteorological Satellite Activities, Weather Bureau, "Policies and Organization," June 1, 1962, reproduced in Rao, *Evolution of the Weather Satellite Program in the US Department of Commerce*, 33–35.

27. To discipline, the system project managers used a sophisticated system of information flow to enable all to keep informed on the status of the program. See John Krige, "Crossing the Interface from R&D to Operational Use," *Technology and Culture* 41 (January 2000); Glenn Seaborg, "Science, Technology, and Development: A New World Outlook," *Science* 181 (July 1973): 13–19.

28. Angelina Long "Making the Atmospheric Science Global: Satellite Development, 'Data-Sparse Regions,' and the World Weather Watch," unpublished paper delivered at American Society for Environmental History, April 2010, 16–17, copy in possession of author.

29. On the IGY, see Roger D. Launius, James Rodger Fleming and David H. DeVorkin, ed., *Globalizing Polar Science: Reconsidering the International Polar and Geophysical Years* (New York: Palgrave Macmillan, 2010); Clark Miller and Paul N. Edwards, ed., *Changing the Atmosphere: Expert Knowledge and Environmental Governance* (Cambridge, MA: MIT Press, 2001).

30. See Homer E. Newell, *Beyond the Atmosphere: Early Years of Space Science* (Washington, DC: NASA SP-4211, 1980), 186–200; John Cloud, "American Cartographic Transformations during the Cold War," *Cartography and Geographic Information Science* 29, no. 3 (2002); *Significant Achievements in Satellite Geodesy, 1958–1964* (Washington, DC: NASA SP-0094, 1966); Deborah Warner, "From Tallahassee to Timbuktu: Cold War Efforts to Measure Intercontinental Distances," *Historical Studies in the Physical and Biological Sciences* 30, part 2 (2000); Naomi Oreskes, ed., *Plate Tectonics: An Insider's History of the Modern Theory of the Earth* (Boulder, CO: Westview Press, 2003); Donald MacKenzie, *Inventing Accuracy: A Historical Sociology of Nuclear Missile Guidance* (Cambridge, MA: MIT Press, 1990).

31. Newell, *Beyond the Atmosphere*, 334.

32. National Academy of Sciences, Space Science Board, *Science in Space* (Washington, DC: National Academy Press, 1961), chap. 3, p. 1.

33. National Academy of Sciences, Space Science Board, *A Review of Space Research: A Report of the Summer Study Conducted under the Auspices of the Space Science Board of the National Academy of Sciences* (Washington, DC: National Academy of Sciences, Publication 1079, 1962), 5–13.

34. Joseph N. Tatarewicz, *Space Technology and Planetary Astronomy* (Bloomington: Indiana University Press, 1990), 103–104; see also Joseph N. Tatarewicz, "Federal Funding and Planetary Astronomy, 1950–1975: A Case Study," *Social Studies of Science* 16, no. 1 (1986).

35. F.T. Barath, A.H. Barrett, J. Copeland, D.E. Jones, and A.E. Lilley, "Mariner 2 Microwave Radiometer Experiment Results," *Astronomical Journal* 69 (1964): 49–58; D.H. Staelin, A.H. Barrett, and B.R. Kusse, "Observations of Venus, the Sun, Moon and Tau A at 1.18-cm Wavelength," *Astronomical Journal* 69 (1964): 69–71; A.H. Barrett and D.H. Staelin, "Radio Observations of Venus and the Interpretations," *Space Science Reviews* 3 (1964); D.H. Staelin and A.H. Barrett, "Spectral Observations of Venus Near 1-centimeter Wavelength," *The Astrophysical Journal* 144(1) (1966).

36. Glenn Fowler, "Dr. Alan H. Barrett, 64, Physicist and Pioneer in Radio Astronomy," *New York Times,* July 4, 1991.

37. Ibid.

38. Staelin published 165 papers between 1964 and 2010, more than two-thirds were on Earth science in one form or another. As examples see D.H. Staelin, "Measurements and Interpretation of the Microwave Spectrum of the Terrestrial Atmosphere near 1-centimeter Wavelength," *Journal of Geophysical Research* 71, no. 12 (1966); D.H. Staelin, F.T. Barath, J.C. Blinn III, E.J. Johnston, "Section 7: The Nimbus E Microwave Spectrometer (NEMS) Experiment," *Nimbus-5 Users Guide* (1972); D.H. Staelin et al., "Microwave Spectrometer on the Nimbus 5 Satellite: Meteorological and Geophysical Data," *Science* 182 (1973): 1339–41; C. Surussavadee and D. H. Staelin, "Global Precipitation Retrievals Using the NOAA/AMSU Millimeter-Wave Channels: Comparisons with Rain Gauges," *Journal of Applied Meteorology and Climate* 49 (2010).

39. Conway B. Leovy, "Simple Models of Thermally Driven Mesospheric Circulation," *Journal of Atmospheric Sciences* 21 (1964); Varavut Limpasuvan, Conway B. Leovy, and Yvan J. Orsolini, "Observed Temperature Two-day Wave and its Relatives Near the Stratopause," *Journal of Atmospheric Sciences* 57 (2000); "Conway B. Leovy: Emeritus Professor of Atmospheric Sciences and Geophysics, July 16, 1933–July 9, 2011," accessed 6/8/2012 12:02 PM. http://www.atmos.washington. edu/people/leovy.shtml.

40. Hugh R. Anderson and P.D. Hudson, "Non-Uniformity of Solar Protons Over the Polar Caps on March 24, 1966," *Journal of Geophysical Research* 74 (1969).

41. Conway Snyder et al., "Interplanetary Space Physics," in *Lunar and Planetary Sciences in Space Exploration* (Washington, DC: NASA SP-14, 1962), 164.

42. S.H. Gross, W.E. Mc Govern, and S.I. Rasool, "The Upper Atmosphere of Mars," NASA-TM-X-57327, 1965; S.H. Gross, W.E. Mc Govern, and S.I. Rasool, "The Atmosphere of Mercury," NASA-TM-X-57322, 1966;

S.H. Gross, W.E. Mc Govern, and S.I. Rasool, "On the Exospheric Temperature of Venus," NASA-TM-X-61176, 1967; J.S. Hogan, R.W. Stewart, S.I. Rasool, and L.H. Russell, "Results of the Mariner 6 And 7 Mars Occultation Experiments," NASA Technical Note TN-D-6683, 1972.

43. Among many others see S.I. Rasool, "Evolution of the Earth's Atmosphere," NASA-TM-X-61176, 1967; S.I. Rasool, "Predicting Earth's Dynamic Changes," *Aerospace America* 24 (January 1986).

44. S.I. Rasool, "Space Observations for Global Change," *Physical Measurements and Signatures in Remote Sensing*, 2 (1991): 805.

45. S.I. Rasool and S.H. Schneider, "Atmospheric Carbon Dioxide and Aerosols: Effects of Large Increases on Global Climate," *Science* 173 (July 9, 1971).

46. Victor Cohn, "U.S. Scientist Sees New Ice Age Coming," *Washington Post*, July 9, 1971, p. A4.

47. George Monbiot, "A Hard Rain's A-gonna Fall," *The Guardian*, May 14, 2004. The book by talk show host Art Bell and Whitley Strieber, *The Coming Global Superstorm* (New York: Pocket Books, 1999), served as the immediate basis for the film.

48. Carl Sagan's accomplishments are too numerous to mention. Key books about him include Keay Davidson, *Carl Sagan: A Life* (New York: John Wiley & Sons, 1999); Tom Head, ed., *Conversations with Carl Sagan* (Jackson: University of Mississippi Press, 2005); William Poundstone, *Carl Sagan: A Life in the Cosmos* (New York: Henry Holt & Company, 1999). He especially applied his knowledge of planetary atmospheres, in this case studies of Venus, to Earth coming up with the "nuclear winter" scenario for climate change in the aftermath of a nuclear exchange between the USA and the Soviet Union. See R. P. Turco et al., "Nuclear Winter: Global Consequences of Multiple Nuclear Explosions," *Science* (December 23, 1983); also Paul R. Ehrlich, Carl Sagan, Donald Kennedy, and Walter Orr Roberts, *The Cold and the Dark: The World after Nuclear War* (New York: W.W. Norton and Co., 1984), 83–85; Lawrence Badash, "Nuclear Winter: Scientists in the Political Arena," *Physics in Perspective* (2001).

49. Eric Nagourney, "Robert Sharp Dies at 92; Linked Study of Planets," *New York Times*, June 14, 2004.

50. At the time, Newell was Superintendent of the Atmosphere and Astrophysics Division, Naval Research Laboratory (NRL). NRL was the organization in charge of the Vanguard Program. Newell and Silverstein had been discussing the transfer of NRL scientists for sometime before NASA opened its doors.

51. NASA, *Policy on Space Flight Experiments* (Washington, DC: NASA, December 12, 1958); Hunley, *The Birth of NASA*, 6–15.

52. Robert L. Rosholt, *An Administrative History of NASA, 1958–1963* (Washington, DC: NASA SP-4101, 1966), 217–26.
53. W. Henry Lambright, *Governing Science and Technology* (New York: Oxford University Press, 1976), 196.
54. Conway, "Earth Science and Planetary Science," 563–86.
55. Pamela E. Mack, *Viewing the Earth: The Social Construction of the Landsat Satellite System* (Cambridge, MA: MIT Press, 1990).
56. The Ozone War was the title of the first popular history of the campaign against CFC use. See Lydia Dotto and Harold Schiff, *The Ozone War* (NY: Doubleday, 1978).
57. James C. Fletcher to Robert Frosch, "Problems and Opportunities at NASA," 9 May 1977, James C. Fletcher Chronological Files, 1977, NASA Historical Reference Collection.
58. This has remained the case thereafter. The US government spent $1.86 billion dollars on climate science in fiscal year 2005; NASA's share of this was a whopping $1.24 billion. The split over technology and science for Earth investigations also remained. NASA spent $519 million on climate-related research, and $722 million on developing and procuring new climate-related space-based observation technologies. The next largest funder of climate science, the National Science Foundation, spent $198 million that year. The National Oceanic and Atmospheric Administration spent only $120 million. See Funding for the Climate Change Science Program, accessed 8/20/2012, 10:26 AM, http://www.usgcrp.gov/usgcrp/Library/ocp2007/default.htm.
59. Goward and Williams, "Landsat and Earth Systems Science: Development of Terrestrial Monitoring," 896.
60. Sally Ride, et al., *NASA Leadership and America's Future in Space: A Report to the Administrator* (Washington, DC: NASA, 1987), accessed 11/16/2014 11:54 AM, http://history.nasa.gov/riderep/main.PDF.
61. Diane E. Wickland, "Mission to Planet Earth: The Ecological Perspective," *Ecology* 72 (December 1991): 1923–33; U.S. Congress, Office of Technology Assessment, *The Future of Remote Sensing From Space: Civilian Satellite Systems and Applications* (Washington, DC: OTA-ISC-558, July 1993), 65–72.
62. Committee on the Assessment of NASA's Earth Science Program, Space Studies Board, *Earth Science and Applications from Space: A Midterm Assessment of NASA's Implementation of the Decadal Survey* (Washington, DC: National Academies Press, 2012), 1–14; NASA, "Earth System Science," 2006, accessed 11/16/2014 12:10 PM, http://www.nasa.gov/pdf/55396main_13%20ESS.pdf.

Acknowledgments The author wishes to thank Jonathan Cohen, Brian Jirout, Amanda Peacock, and Megan Porter for research assistance on this essay, as well as Erik Conway and Monique Laney who commented on drafts.

8.6   PHOTOGRAPHS

Photograph 8.1   Homer E. Newell (L), NASA Associate Administrator for space science, in 1962. Newell was responsible for initiating and developing the science program at the new space agency (NASA Photo, public domain)

Photograph 8.2   A photo map of the contiguous 48 states of the USA was the first ever assembled from satellite images. It was completed in 1974 for NASA by the US Department of Agriculture Soil Conservation Service Cartographic Division, measured 10 by 16 ft, and is composed of 595 cloud-free black-and-white images returned from NASA's first Earth Resources Technology Satellite (ERTS-1) (*Credit:* NASA, public domain, available on-line at http://grin.hq.nasa.gov/ABSTRACTS/GPN-2003-00031.html)

Photograph 8.3   From the past to the present: global view of Earth produced using imagery acquired in 2001 (*Credit:* NASA Goddard Space Flight Center Image by Reto Stöckli (land surface, shallow water, clouds). Enhancements by Robert Simmon (ocean color, compositing, 3D globes, animation) Public domain (Available on-line at http://visibleearth.nasa.gov/view.php?id=57723))

References

Anderson, Hugh R., and P.D. Hudson. 1969. Non-uniformity of solar protons over the polar caps on March 24, 1966. *Journal of Geophysical Research* 74: 2881–2890.

Badash, Lawrence. 2001. Nuclear winter: Scientists in the political arena. *Physics in Perspective* 3: 76–105.

Barath, F.T., A.H. Barrett, J. Copeland, D.E. Jones, and A.E. Lilley. 1964. Mariner 2 microwave radiometer experiment results. *Astronomical Journal* 69: 49–58.

Barrett, A.H., and D.H. Staelin. 1964. Radio observations of Venus and the interpretations. *Space Science Reviews* 3: 109–135.

Bates, Charles C., and John F. Fuller. 1986. *America's weather warriors, 1814–1985*. College Station: Texas A&M University Press.

Bell, Art, and Whitley Strieber. 1999. *The coming global superstorm*. New York: Pocket Books.

Boix-Mansilla, Veronica. 2010. Learning to synthesize: The development of interdisciplinary understanding. In *The Oxford handbook of interdisciplinarity*, ed. Robert Frodeman, Julie Thompson Klein, Carl Mitcham, and J. Britt Holbrook, 288–292. New York: Oxford University Press.

Borisenkov, Y.P. 1974. Global Atmospheric Research Program (GARP). *Meteorology and Hydrology* 12: 155–162.

Cloud, John. 1966. *Significant achievements in satellite geodesy, 1958–1964*. Washington, DC: NASA SP-0094.

Cloud, John. 2002. American cartographic transformations during the Cold War. *Cartography and Geographic Information Science* 29(3): 261–282.

Cohn, Victor. 1971. U.S. scientist sees new ice age coming. *Washington Post*, July 9, p. A4.

Collins, Harry, Robert Evans, and Mike Gorman. 2007. Trading zones and interactional expertise. *Studies in History and Philosophy of Science Part A* 38(4): 657–666.

Committee on the Assessment of NASA's Earth Science Program, Space Studies Board. 2012. *Earth science and applications from space: A midterm assessment of NASA's implementation of the decadal survey*. Washington, DC: National Academies Press.

Conway, Erik M. 2008. *Atmospheric sciences at NASA: A history*. Baltimore: Johns Hopkins University Press.

Conway, Erik M. 2010. Earth science and planetary science: A symbiotic relationship? In *NASA's first 50 years: Historical perspectives*, ed. Steven J. Dick, 563–586. Washington, DC: NASA SP-2010-4070.

Crouch, Tom D. 1983. *The eagle aloft: Two centuries of the balloon in America*. Washington, DC: Smithsonian Institution Press.

Cullen, Heidi. 2012. Clouded forecast. *New York Times*, May 31.

Davidson, Keay. 1999. *Carl Sagan: A life*. New York: Wiley.

Davies, Merton E., and William R. Harris. 1988. *RAND's role in the evolution of balloon and satellite observation systems and related U.S. space technology*. Santa Monica: The RAND Corporation.

de Syon, Guilllaume. 2002. *Zeppelin! Germany and the airship, 1900–1939*. Baltimore: Johns Hopkins University Press.

Dotto, Lydia, and Harold Schiff. 1978. *The ozone war*. New York: Doubleday.

Ehrlich, Paul R., Carl Sagan, Donald Kennedy, and Walter Orr Roberts. 1984. *The cold and the dark: The world after nuclear war*. New York: W.W. Norton and Co.

Forman, Paul. 1987. Behind quantum electronics: National security as basis for physical research in the United States, 1940–1960. *Historical Studies in the Physical and Biological Sciences* 18(1): 149–229.

Fowler, Glenn. 1991.Dr. Alan H. Barrett, 64, physicist and pioneer in radio astronomy. *New York Times*, July 4.

Gavaghan, Helen. 1998. *Something new under the Sun: Satellites and the beginning of the space age*. New York: Copernicus Books.

Goward, Samuel N., and Darrel L. Williams. 1997. Landsat and earth systems science: Development of terrestrial monitoring. *Photogrammetric Engineering & Remote Sensing* 63: 887–900.

Head, Tom (ed.). 2005. *Conversations with Carl Sagan*. Jackson: University of Mississippi Press.

Hill, Janice. 1991. *Weather from above*. Washington, DC: Smithsonian Institution Press.

Hunley, J.D. (ed.). 1993. *The birth of NASA: The diary of T. Keith Glennan*. Washington, DC: NASA SP-4105.

Krige, John. 2000. Crossing the interface from R&D to operational use. *Technology and Culture* 41: 27–50.

Lambright, W. Henry. 1976. *Governing science and technology*. New York: Oxford University Press.

Launius, Roger D. 2012. Imprisoned in a tesseract: NASA's human spaceflight effort and the prestige trap. *Astropolitics: The International Journal of Space Politics and Policy* 10(2): 152–175.

Launius, Roger D., James Rodger Fleming, and David H. DeVorkin (eds.). 2010. *Globalizing polar science: Reconsidering the International Polar and Geophysical years*. New York: Palgrave Macmillan.

Leovy, Conway B. 1964. Simple models of thermally driven mesospheric circulation. *Journal of Atmospheric Sciences* 21: 327–341.

Limpasuvan, Varavut, Conway B. Leovy, and Yvan J. Orsolini. 2000. Observed temperature two-day wave and its relatives near the stratopause. *Journal of Atmospheric Sciences* 57: 1689–1701.

Mack, Pamela E. 1990. *Viewing the Earth: The social construction of the Landsat satellite system*. Cambridge, MA: MIT Press.

MacKenzie, Donald. 1990. *Inventing accuracy: A historical sociology of nuclear missile guidance*. Cambridge, MA: MIT Press.

McQuaid, Kim. 2006. Selling the space age: NASA and Earth's environment, 1958–1990. *Environment and History* 12: 127–163.

Miller, Clark, and Paul N. Edwards (eds.). 2001. *Changing the atmosphere: Expert knowledge and environmental governance*. Cambridge, MA: MIT Press.

Monbiot, George. 2004. A hard rain's a-gonna fall. *The Guardian*, May 14.

Morrow Jr., John H. 1993. *The Great War in the air: Military aviation from 1909–1921*. Washington, DC: Smithsonian Institution Press.

Nagourney, Eric. 2004. Robert Sharp dies at 92; Linked study of planets. *New York Times*, June 14.

NASA. 1958. *Policy on space flight experiments*. Washington, DC: NASA.

National Academy of Sciences, Space Science Board. 1961. *Science in space*. Washington, DC: National Academy Press.

National Academy of Sciences, Space Science Board. 1962. *A review of space research: A report of the summer study conducted under the auspices of the Space Science Board of the National Academy of Sciences*. Washington, DC: National Academy of Sciences, Publication 1079.

National Research Council. 1969. National Academy of Sciences annual report, fiscal year 1968–1969.

Nebeker, Frederik. 1995. *Calculating the weather: Meteorology in the 20th century*. San Diego: Academic Press.

Newell, Homer E. 1980. *Beyond the atmosphere: Early years of space science*. Washington, DC: NASA SP-4211.

Nkemdirim, L.C. 1975. The Global Atmospheric Research Program and the geographer. *The Professional Geographer* 27: 227–230.

Oreskes, Naomi (ed.). 2003. *Plate tectonics: An insider's history of the modern theory of the Earth*. Boulder: Westview Press.

Polavarapu, R.J., and G.L. Austin. 1979. A review of the GARP Atlantic Tropical Experiment (GATE). *Atmosphere–Ocean* 17(1): 2–13.

Poundstone, William. 1999. *Carl Sagan: A life in the cosmos*. New York: Henry Holt & Company.

Rao, P. Krishna. 2001. *Evolution of the weather satellite program in the U.S. Department of Commerce: A brief outline*. Washington, DC: NOAA.

Rasool, S.I. 1986. Predicting Earth's dynamic changes. *Aerospace America* 24: 78–80, 82.

Rasool, S.I. 1991. Space observations for global change. In *ESA: Physical measurements and Signatures in remote sensing*. Abstract available on-line at http://ntrs.nasa.gov/search.jsp?R=19920002337, accessed 5/16/2016. vol. 2, 805.

Rasool, S.I., and S.H. Schneider. 1971. Atmospheric carbon dioxide and aerosols: Effects of large increases on global climate. *Science* 173: 138–141.

Ride, Sally, et al. 1987. *NASA leadership and America's future in space: A report to the administrator.* Washington, DC: NASA.

Rolt, L.T.C. 1966. *The aeronauts: A history of ballooning—1783–1903.* New York: Walker and Company.

Rosholt, Robert L. 1966. *An administrative history of NASA, 1958–1963.* Washington, DC: NASA SP-4101.

Sachs, Wolfgang. 1999. *Planet dialectics: Explorations in environment and development.* London: Zed Books.

Seaborg, Glenn. 1973. Science, technology, and development: A new world outlook. *Science* 181: 13–19.

Shikazono, Naotatsu. 2012. *Introduction to Earth and planetary system science: New view of Earth, planets, and humans.* New York: Springer.

Snyder, Conway, Hugh R. Anderson, Marcia Neugebauer, and Edward J. Smith. 1962. Interplanetary space physics. In *Lunar and planetary sciences in space exploration.* Washington, DC: NASA SP-14.

Staelin, D.H. 1966. Measurements and interpretation of the microwave spectrum of the terrestrial atmosphere near 1-centimeter wavelength. *Journal of Geophysical Research* 71(12): 2875–2881.

Staelin, D.H., and A.H. Barrett. 1966. Spectral observations of Venus near 1-centimeter wavelength. *The Astrophysical Journal* 144(1): 352–363.

Staelin, D.H., A.H. Barrett, and B.R. Kusse. 1964. Observations of Venus, the Sun, Moon and Tau A at 1.18-cm wavelength. *Astronomical Journal* 69: 69–71.

Staelin, D.H., F.T. Barath, J.C. Blinn III, and E.J. Johnston. 1972. Section 7: The Nimbus E microwave spectrometer (NEMS) experiment. In *Nimbus-5 users guide,* ed. Goddard Space Flight Center, 141–162. Baltimore: Allied Research Associates.

Staelin, D.H., A.H. Barrett, J.W. Waters, F.T. Barath, E.J. Johnston, P.W. Rosenkranz, N.E. Gaut, and W.B. Lenoir. 1973. Microwave spectrometer on the Nimbus 5 satellite: Meteorological and geophysical data. *Science* 182: 1339–1341.

Stevens, William K. 1999. *The change in the weather: People, weather, and the science of climate.* New York: Delacorte Press.

Surussavadee, C., and D.H. Staelin. 2010. Global precipitation retrievals using the NOAA/AMSU millimeter-wave channels: Comparisons with rain gauges. *Journal of Applied Meteorology and Climate* 49: 124–135.

Tararewicz, Joseph N. 1986. Federal funding and planetary astronomy, 1950–1975: A case study. *Social Studies of Science* 16(1): 79–103.

Tatarewicz, Joseph N. 1990. *Space technology and planetary astronomy.* Bloomington: Indiana University Press.

Turco, R.P., O.B. Toon, T.P. Ackerman, J.B. Pollack, and Carl Sagan. 1983. Nuclear winter: Global consequences of multiple nuclear explosions. *Science* 222: 1283–1292.

U.S. Congress, Office of Technology Assessment. 1993. *The future of remote sensing from space: Civilian satellite systems and applications.* Washington, DC: OTA-ISC-558.

U.S. National Coordinating Committee for Aviation Meteorology, Panel of Operational Meteorological Satellites. 1961. *Plan for a National Operational Meteorological Satellite System.* Washington, DC: U.S. Government Printing Office.

Warner, Deborah. 2000. From Tallahassee to Timbuktu: Cold War efforts to measure intercontinental distances. *Historical Studies in the Physical and Biological Sciences* 30 (part 2): 393–415.

Wexler, Harry. 1954. Observing the weather from a satellite vehicle. *Journal of the Interplanetary Society* 13: 269–276.

Wickland, Diane E. 1991. Mission to planet Earth: The ecological perspective. *Ecology* 72: 1923–1933.

Witkins, Richard. 1960. U.S. orbits weather satellite; It televises Earth and clouds; New era in meteorology seen. *New York Times*, Apr 2, p. 1.

CHAPTER 9

Interdisciplinary Research and Transformative Research as Facets of National Science Policy

Irwin Feller

9.1   INTRODUCTION

"As science evolves, how can science policy?" a question posed by Benjamin Jones,[1] is a science of science policy-focused statement of the key themes articulated by the editors in the introduction to this book. For permeating and overlapping discourse on how new scientific fields are formed are questions on how to organize and fund scientific renewal that constitute policy issues for decision makers within and across funding agencies and universities. This chapter focuses on two issues: interdisciplinary research and transformative research, nested within the larger set of national science policy questions. For purposes of exposition, it treats each issue separately to bypass debates surrounding the extent to which interdisciplinary approaches are either necessary or sufficient for transformative research.[2]

I. Feller (✉)
Pennsylvania State University, University Park, PA, USA

© The Editor(s) (if applicable) and The Author(s) 2016   243
T. Heinze, R. Münch (eds.), *Innovation in Science and
Organizational Renewal*, DOI 10.1057/978-1-137-59420-4_9

The chapter takes as given widely reported developments in interdisciplinarity, as, for example, in the physical, biological, and other sciences.[3] Instead, its focus is on the interdependent decisions that universities and government funding agencies are making and purportedly need to make to accommodate and/or nurture asserted puissant trends in the performance of academic scientific research. From this perspective, calls for interdisciplinarity and transformative research are essentially variants of long-standing, oft-encountered science policy questions, succinctly inventoried in the Office of Technology Assessment's (OTA) 1991 report: "Federally Funded Research: Decisions for a Decade," about which fields of science to fund, which actors (individuals/teams/institutions) are best qualified to conduct the chosen fields of inquiry, what mechanisms should be used to select performers and funding mechanisms, and, finally, what criteria or methodologies should be used to assess the quantity, quality, or relevance of the ensuing research findings.[4] Overshadowing and conditioning answer to these questions are the perennial practical constraints of resource availability. Phrased more pointedly, the total resources required to satisfy the claims for continued support of "mainstream" research fields (or disciplines), to underwrite the reconfigurations of these fields about new interdisciplinary paradigms, and to meet the claims of advocates for support of new fields directed at societal problems, for example, "vulnerability science,"[5] will push up against and invariably exceed whatever level of total resources are provided for by the collectivity of sponsors—governments, foundations, industry, universities—for this research.

This is not a novel statement. It is a generic formulation applicable widely across Organisation for Economic Co-operation and Development (OECD) members. The specific ways though in which these questions are answered as well as the content of the answers themselves are context dependent, shaped by specific, historic, path-dependent events, or what Weiller has termed "national structural preferences regarding the utilization and construction of the productive forces within each country."[6] The national context within which issues related to interdisciplinary research and transformative research are examined in this chapter is that of the USA. This immediately brings to the fore a national research system in which national government support of basic research is channeled primarily to academic universities, and where this funding is allocated in the main via competitive, merit review procedures, historically rooted in single investigator-initiated funding. While there are manifest similarities in conditions between the USA and other OECD countries in the trends reported on below,[7] there also are differences[8]; thus the form and fit of the policy analyses presented here to other national contexts has yet to be determined.

9.2 INTERDEPENDENCE

The analysis below moves back and forth between changes in funding agencies and in universities in order to highlight the interdependence of actions between the two sectors. The approach though understates the historic and contemporary roles of foundations, such as Rockefeller, Ford, and Howard Hughes, and of firms, as in the information technology and pharmaceutical industries, as prime movers and/or catalysts in engendering the capacities of universities to explore new research fields, thereby setting into motion the subsequent sequence of interdependent steps between universities and governments relating to interdisciplinary and transformative research.[9] It is however useful in highlighting nuanced differences in the role of the two sectors as between initiatives to foster interdisciplinary research and transformative research.

Focusing initially on interdisciplinarity, much of the impetus underlying its emergence on American campuses has come from the Federal government.[10] The establishment of academic programs and research centers devoted to materials science, by now a ubiquitous, well-established field, exemplified by the National Science Foundation's (NSF) Materials Research and Science and Engineering Centers program, is but one example of these dynamics. According to Harwood,[11] a participant in the formative years of the field, the impetus for prodding universities to reconfigure their research and teaching approaches to materials reflected in good part the "fantastic set ... of problems and challenges" faced by government agencies for new materials that could resist higher stresses, temperatures, pressures, aggressive environments, and so on, found in new weapons systems, propulsion systems, electronic devices, and more.[12] This led to substantial support first by the Defense Department's Advanced Research Projects Agency (DARPA), and then to large-scale, continuing support by the NSF. According to several participants in the rise of this field, external support was an essential element in providing the discretionary resources and external legitimacy that both encouraged and then subsidized universities to make large-scale investments in faculty and facilities to accommodate materials science, as well as to make the organizational changes needed to administer these new programs.

More generally, noting the broader trends since the 1970s toward interdisciplinarity, most noticeably in the natural sciences and engineering, Miller has observed that "The primary driver for this shift towards interdisciplinary in the natural sciences and engineering was large-scale government funding of research oriented toward the solution of societal problems, from curing disease and ensuring national security to protecting the environment."[13] But agencies, as with universities can experience internal contests between established disciplinary-based divisions and newer cross-disciplinary programs. Priority setting both among and within fields of science is a well recognized perennial practical problem in science policy, with the search for a science of science policy that would provide grounded decision rules a continuing if still elusive objective.[14] Additionally, in an open science system redolent of a republic of science, wherein representatives of the academic community serve on the advisory and policy-making bodies of funding agencies, and as short-term, rotating program managers, helping to shape agency research priorities as well as recommending and making funding decisions, faculty carry messages about the saliency of modes of research in both directions, causing intra-agency and intra-university contests about relative emphases to become intertwined.

9.3   ANALYTICAL FRAMEWORK AND POLICY CONTEXT

The chapter's starting points as noted above are oft-made statements that the structural dynamics of scientific inquiry have changed in the direction of requiring increased collaboration across traditional disciplinary boundaries as well as those that assert that new initiatives are needed to jump start the overly conservative, incremental pace at which scientific knowledge is advancing, or being transformed. Likewise, it acknowledges, and indeed in other writings[15] has described the portfolio of changes that universities have made to foster and/or to accommodate at least a modicum of interdisciplinary research (and graduate education).

In analytical style and tone, though, it is more guarded about the strength or the structural importance of trends toward interdisciplinarity and transformative research than many recent treatments. In the context of the science policy decisions it considers, it sees the case for initiatives to bolster interdisciplinary or transformative research as neither unchallengeable or unstoppable. Additionally, assuming that a dominating case is made for either or both of these trends, little consensus exists about the form such initiatives should take, quite the contrary.

Considerable disagreement exists about the degree to which either agencies or universities need to make "fundamental" or "transformative" adjustments to accommodate either or both of above. The disagreements occur both at the conceptual level, especially about the construct validity of what is meant by transformative research, and at policy design and implementation levels about the shape and size of the operational and organizational changes needed within funding agencies or universities to accommodate interdisciplinary or transformative research initiatives.

The disagreements in part reflect long-standing differences about the funding and organizational arrangements deemed most conducive to the pursuit and attainment of advances in scientific knowledge, as for example, in continuing debates about the relative importance in physics of "little" and "big" science.[16] These differences have not gone away. As applied to interdisciplinarity, Weingart's projection that in all likelihood within universities "... traditional and inter-, multi- and transdisciplinary research fields will exist side by side,"[17] is likely accurate but only to a point. Observations pointing to coexistence, or parallel play, found in selected fields of science, colleges, or departments does not negate the tensions and contests that can exist within a university whenever decisions need to be made about how best to organize itself to maintain or enhance its reputational standing for research excellence, including decisions about which academic units are to be maintained, changed or terminated and what faculty expertise needs to be recruited to support these decisions. Relatedly, accounts of successful examples of how universities have transformed themselves to accommodate interdisciplinarity[18] reflect the accumulated effects of past events; they are not necessarily prognoses of future trends. As illustrated by the headline "Hiring that Crosses Disciplines Can Create Tensions" in a recent "Chronicle of Higher Education" report,[19] externally oriented visions and initiatives from senior university administrators about the opportunities and imperatives associated with interdisciplinarity cannot always penetrate the inward-looking, small-world perspective and autonomy of discipline-based departments.[20] Success stories, per se, are akin to sampling on the dependent variable: successful undertakings. Such accounts ignore stillborn or failed endeavors. Moreover, success, even when documented, may represent only the early stages of a specific program not its long-term sustainability, especially if it was launched with a heavy external subsidy. Success also may produce an interdisciplinary enclave that generates little spillover impacts toward a more supportive environment for interdisciplinarity in colleges or departments elsewhere in the university.

Of concern here is that exegesis and case study narrative on interdisciplinarity from an epistemic or historic perspective at times seems to take on normative connotations, implying that future advances in science require even more of a transformation in organizational and funding arrangements than has occurred to date for renewal to occur or accelerate. Whatever may be the case for this position, and in good part it is a strong one given prevailing trends in many countries to focus incremental government research support on "grand challenges," prognoses of the future(s) of interdisciplinary approaches to scientific research do not sufficiently take into account that the conditions that nurtured interdisciplinary research over the past 25 years or so have changed, and in an unfavorable direction.

Two contemporary trends in particular are likely to create a less hospitable environment for future initiatives. The first from which little near-term relief appears forthcoming across OECD nations are the combination of stagnant levels of resources to support new ventures or approaches in scientific research and widely perceived assessments on the part of researchers in many countries that their governments are pressuring them to give priority to shorter-term, more applied research, frequently directed at economic competitiveness.[21] Not only is funding being reallocated in this way, but the pressures for near-term results work against the upfront, often time-consuming efforts needed to build trust among researchers from different disciplines and to acculturate each to the other's discipline, traits deemed essential to sustainable interdisciplinary endeavors.[22]

Focusing again only on the USA and abstracting from the vicissitudes of annual agency research and development (R&D) budgets caused by the dysfunctional character of contemporary US politics, downward pressures on the growth of the US federal budget and upward built-in pressures from entitlement programs and national debt payments have combined to exert downward pressures on the share of non-discretionary appropriations in the Federal budget: that portion of the budget in which R&D competes with all other domestic programs, many highly regarded and/or highly protected in their own right. As a consequence, even though support of basic research is widely accepted as a legitimate government function, enjoying broad-based political support, with the notable exceptions of research areas that raise ideological or interest group hackles—climate research, social science research, civilian technology programs—larger budget pressures have placed de facto caps on the growth of Federal support of R&D. These pressures may be seen in a pronounced long-term decline in the ratio of Federally funded R&D to GDP from approximately 1.2% in 1976 to under 0.8% in 2014.

This macro-level fiscal environment has led in turn to tighter funding conditions for academic researchers, even as total science agency budgets have increased. Set against the continuing growth in the academic scientific enterprise represented by the increase in number of research-intensive and research-oriented universities staffed by a larger numbers of faculty whose careers and aspirations are geared to research, the consequence has been an increase in proposals and requests for funds submitted to funding agencies; the predictable result has been a decrease in success ratios. At the National Institutes of Health (NIH), the single major Federal government agency sponsor of academic R&D, the number of research project grants increased from 40,861 to 49,581 between 2004 and 2013, while the success or funding rates fell from 24.6% in 2004 to 16.8% in 2013. Perhaps even more striking is the success rate for RO1-like grants, the emblematic metric of the USA's reliance on single investigator-initiated research, fell from 25.4% to 17.5% over this period, even as the number of applications reviewed remained relatively unchanged, at approximately 28,000. At NSF, the decline, from 24% to 22%, was muted somewhat by the waxing and waning of proposal submissions, as faculty appeared to adjust rapidly with a one- to two-year lag in NSF's funding and ability to support new proposals.

An inescapable consequence of this tightly constrained resource environment is to limit the flexibility that Federal agencies have to engender renewal of science through cumulative incremental change, whereby additions to agency budgets ("slack resources") are used to add continuously to the total scale and scope of new arrangements until through a process of functional displacement they become the norm for policy and practice, all this however without having to reduce absolute levels of support of what was widely to have worked before.[23] Interdisciplinary research centers, for example, could be funded by science agencies without reducing the absolute level of funding available for RO1/single investigator grants or overly detracting from the resources or disrupting the rhythms of departmental or college life. The new resource environment instead paints choices between existing and new(er) ways of doing things in vivid zero-sum colors, much more likely to foster resistance and opposition to new initiatives, whatever their form. Casting these dynamics in historical terms, increases in the availability of research support have served to mute the jarring that frequently occurs when innovations are introduced into highly routinized organizations. The new funding environment changes all this. If the recent past may be described as one in which a rising tide lifts all boats, the present and near-term future more closely resembles (or is perceived as resembling) Hitchcock's *Lifeboat*.

The second trend are the increased requirements, variously labeled under the headings of neoliberalism, new public management and/or evidence-based decision-making, imposed by governments for accountability and evidence of performance on the part of those receiving public sector support. These trends are widespread across OECD nations across many functional areas. In the USA, these requirements are most often associated with the initial enactment of the Government Performance and Results Act and related policy edicts issued by the Executive branch, Office of Management and Budget, requiring that agencies more systematically evaluate their programs.

Funding of academic research has not escaped these new requirements. Demands for evidence of performance are not neutral with respect to the prospects either for interdisciplinary research or for transformative research. The challenge of demonstrating quality is a staple issue in discussions of interdisciplinary research.[24] The challenges are heightened in this new era. For in a policy environment that requires methodologically rigorous evidence and apotheosizes random control trials, failure to provide such about the impacts of what is new unavoidably favors the default position of pre-existing arrangements. Thus according to Whitley, changes in the governance of science, encompassing in combination "steady state levels of funding, project-based resource allocation and formal, public monitoring of PRO's performance may well restrict intellectual diversity and encourage scientists to work on mainstream topics with established techniques in preference to tackling interdisciplinary problems with a variety of novel methods and approaches that challenge established problems with a variety of methods and approaches that challenge established boundaries and identities."[25] Only recently have such methodologies and findings emerged[26] but they have yet to filter into policy discussions or decision-making settings.

9.4   INTERDISCIPLINARITY AS A COUPLED PROBLEM OF SCIENTIFIC CHOICE AND ACADEMIC GOVERNANCE

Statements about the imperatives of interdisciplinarity pervade contemporary discussion of the dynamics of scientific research. The chorus includes conceptual arguments, as noted above, that a new mode of transdisciplinary knowledge production has emerged alongside the "traditional" disciplinary structure of science and technology[27] as well as expressions about the essentiality of interdisciplinary approaches to research and training by the leaders of NIH and NSF.[28] Calls for an increased emphasis on interdisciplinary approaches to research and teaching have indeed been so widespread in agency documents and reports by prestigious national organizations, such as the National Academies and professional organizations, over the past 25 years or so as early as 1999 they were termed the "mantra of science policy."[29]

As with federal science policy, interdisciplinarity may also be termed the "mantra of academic strategic plans" written since about 2000.[30] The environmental scans routinely included in these plans serve here as guides to the perceptions of universities about the threats and opportunities found in the competitive world they inhabit as they pursue objectives of reputation, ranking, and prestige. Indeed, the very advent and rapid spread of formal strategic planning was itself a mark of the changed external environment.[31] Changes in the rhetoric with which these plans are written also serve indirectly to identify changes in the weights accorded to discipline-based and interdisciplinary initiatives.

The emphasis on interdisciplinarity found in strategic plans written since 2000 stands out in comparison with the thrusts of the strategies of only a few years earlier. At a number of universities, or colleges, or departments, the strategic objective of being among the "best" (or best in class) became identified with upward moves through the Carnegie Classification system and/or placement in the NRC's ranking of doctorate degree programs. The NRC's 1995 rankings, however, encompassed only 41 fields, and these were typically long-established mainstream disciplines. The combination of two presumably unassailable objectives: (1) improved performance/excellence, and (2) more efficient use of resources, tended to skew hard resource allocation choices toward the status quo of pre-existing discipline-based departments.

Contributing to this tendency was the decentralized nature of the university, in which colleges were often left with considerable discretion to emphasize disciplinary-based strategies within the broad but relatively loose guidance of university-level pronouncements about the importance of interdisciplinarity.[32] The result, according to the 1994 report of the Government-University-Industry Research Roundtable (GUIRR) was that interdisciplinary programs became "... 'orphans' within the fiscal bureaucracy of the university. These programs are at a further disadvantage since most of the university's planning efforts are based on fiscal structure. Thus, interdisciplinary programs play a less prominent role in the long-range planning of the university."[33]

The post-2000 plans convey a more supportive commitment to inter-disciplinarity. Echoing the themes found in the national forums outlined above, the environmental scans contained in these plans highlighted changes in both the internal dynamics of science and the external funding environment. More selectively, at a tactical level, the plans highlighted perceived opportunities for the specific university to exploit existing academic strengths or to concentrate modest strengths to carve out distinctive research or graduate degree niches in emerging, often interdisciplinary fields before other institutions enter or preempt them.

Duke University's 2001 strategic plan, "Building on excellence," is cited here as a illustrative example of several other university plans, chosen in part for its articulateness but also because of extent to which implementation followed strategy, with seven campus-wide research institutes embodying the plan's emphasis on interdisciplinarity being created: "While the modern research university was forged from an alliance of disciplines, with knowledge largely fostered within traditional departmental or school structures, recent decades have seen an accelerated integration of knowledge across the sciences, social sciences and humanities, in fields ranging from the biosciences to cultural studies. The mode of research that permits this integration of knowledge can be characterized, to a substantial degree, as multidisciplinary and interdisciplinary."[34] The plan is noteworthy also for its acknowledgment of a need to lower or overcome barriers to multidisciplinary and interdisciplinary activities: "Our task is to facilitate interdisciplinary research and training through incentives, encouragement, rewards, and the removal of institutional, bureaucratic and intellectual barriers."[35]

Recent academic strategic plans convey a more knife-edged equilibrium between discipline-based and interdisciplinary initiatives, along the lines of Weingart's observation cited above. As an example, the University of Wisconsin's (UoW) relatively recent plan: "A Strategic Framework the University of Wisconsin-Madison, 2009–2014" speaks both of continuing to "invest in interdisciplinary life science and biotechnology, including the scientific and engineering disciplines that support twenty-first century biology," but also of acting to "Ensure strength in the core disciplines while promoting innovation, interdisciplinary connections, where it makes intellectual sense to do so."[36] The language however is so general as to obviate any commitment to the actions needed to implement the strategy, as provided for in the Duke plan.

Abstracting from considerations about the extent to which actions described in strategic plans have been implemented, or if implemented, successful, are beyond the scope of this chapter,[37] it still needs to be recognized that these calls have not gone unchallenged. Debate about the relative importance of disciplinary and interdisciplinary modes of research and teaching is lively and continuing. Included here are strong statements about the enduring nature of disciplines, and thus the cyclical passing faddishness of interdisciplinarity.[38]

Disciplines continue to be seen as the most effective and efficient manner for generating scientific and technological advances, training the next generation of scientists, and transferring received knowledge. As contended by Abbott: "Because of their extraordinary ability to organize in one single structure research fields, individual careers, faculty hiring and undergraduate education, disciplinary departments are the essential and irreplaceable building blocks of American universities."[39] Further, "... if interdisciplinarity were going to reorganize the university it would have done so long ago ...."[40]

Cogent arguments based on theories of scientific progress and efficient academic administration support these arrangements. As noted by James Conant at Harvard's Tercentenary celebration in 1936: "... it is because of specialization that knowledge advances, not in spite of it; and that cross fertilization of ideas is possible only when new ideas arise through the intense cultivation of special fields."[41] Similarly, commenting on the unfavorable press that disciplines have received for some time now, Servos has argued that the "discipline not only confines, it also liberates ... The discipline is not dysfunctional; it is functional ... Disciplines not only lend structure and meaning to lives, they also bring order and significance to knowledge."[42] Firming the theoretical base for these positions is the observation of George Stigler, University of Chicago economist, that "specialism is the royal road to efficiency in intellectual as in economic life."[43] Indeed, although the advent of interdisciplinary approaches to scientific research is often presented as a necessary reconfiguration of existing disciplines to advance knowledge, at times creating a new field that displaces earlier ones, it also is possible for recently emerging fields to begin as broad interdisciplinary initiatives but over time segment into a small number of "specialities, or quasi-disciplines, as appears to have happened in 2014 to Duke University's erstwhile campus-wide Duke Institute of

Genome Sciences & Policy,[44] or even to fragment along disciplinary fault lines, as reported by Raasch et. al.[45] in the case of open source innovation.

Rhoten's study of researchers in 13 NSF-funded interdisciplinary centers for environmental research arrives at much the same conclusion: "While the findings suggest that such a transformation toward interdisciplinary research is in fact underway in the centers we have examined, we also conclude from this small sample that, like other recent studies have found in Europe and the United States, the metamorphosis toward interdisciplinary collaboration is less prevalent and progressive than some analysts speculate."[46] As observed by Weingart, following his description of avowedly interdisciplinary fields such as climate research and gender studies joined together in research centers and funding programs but intellectually independent and developing individually: "The replacement of discipline-based mode of knowledge is not corroborated by empirical data either."[47]

For all the emphasis given to interdisciplinarity in strategic plans: "Academic disciplines continue to dominate the modern university, developing curriculum, marshaling resources, administering programs, and doling out rewards."[48] Recent strategic plans continue to emphasize disciplinary-based paths to increased institutional prominence. Interdisciplinarity, for the most part, is presented as a boundary-crossing trend that offers new opportunities, and as a force to which the university must respond; it is not presented, however, as a new dominant intellectual or organizational paradigm.[49] Few plans venture a fundamental restructuring of existing college and departmental structures to accommodate or accelerate a transition to interdisciplinary modes of research and instruction.

But beyond defenses rooted in epistemological theories of scientific progress or the socialization role of professions, there are sound pragmatic reasons for a university to select a disciplinary-based strategy of institutional advancement. As illustrated by the rapid ascendancy to national prominence of New York University's philosophy department, the strategy can work. Indeed, if anything, the NYU experience illustrates the reputational benefits that can be garnered not only by selecting a discipline as the path of advance, but even of a greater concentration of emphasis and resources into a subdiscipline, in this case analytical philosophy.[50]

Yet another retarding factor on the further diffusion of interdisciplinary research is the lack of faculty interest in these initiatives, especially among senior faculty who have well-established and presumably well-funded ongoing disciplinary-based research programs. Thus, for example, posing the question of whether or not former Harvard President Rudenstine's initiatives to foster interdisciplinarity would take hold, Keller and Keller commented in their history of the university that: "It is no means clear that if faculty horses are enticed to gather at center-selected waterholes, they will drink deeply."[51] Again, without the participation of well-established faculty to lend their reputations to new initiatives (as well as experience in conceptualizing and organizing complex research or educational undertakings), interdisciplinary initiatives may fail for want of the right or effective leaders.

9.5 TRANSFORMATIVE RESEARCH

Analyzing the place of transformative research in the constellation of new initiatives to create scientific institutional renewal, again from the perspective of science policy, involves a different constellation of actors, interests, and events. The starting points here are (1) the widespread international bandwagon interest in transformative/breakthrough/high-risk research, as represented by the 2007 National Science Board report: "Enhancing Support of Transformative Research" at the NSF, and (2) terminological and analytic deconstructions that highlight first the conceptual fuzziness of the term and next the programmatic and empirical challenges of both ex ante identification of transformative lines of research (or researchers) and ex post assessment (and acceptance) that transformative findings have been produced.

A classic movie line, especially for devotees of Leslie Howard's 1935 performance as the Scarlet Pimpernel, is: "They seek him here, they seek him there, those Frenchies seek him everywhere. Is he in heaven or is he in hell? That damned elusive Pimpernel." Much the same has been said about the surge of interest in transformative research. Lal in a recent survey of international policy initiatives to support transformative research has argued that "... the concept of transformative research remains mired in mystique. Its definitions tend to be sublime-inspirational but vague. More importantly, as the diversity of program implementation shows ... there is no operational understanding of how to *prospectively* identify transformative research."[52] Similarly, Dietz and Rogers have observed that the NSF definition of transformative research neither emerges "from the empirical investigation of scientific and technological research as phenomena,"[53] nor has a definition been settled on.

The dominant contemporary lines of inquiry about transformative research have focused on the ability of researchers or policy makers to correctly identify and/or predict the lines of research or the researchers who are most likely to generate such findings; thus the insertion of "potentially" before transformative research in subsequent NSF program announcements. There is good reason for this attention.

The arenas of contestation surrounding transformative research for any other than incremental modifications in existing funding patterns or institutional arrangements are likely to be predictability and assessment. The predictability contest is likely to set in opposition histories of science that point to the elusiveness of predictions about future lines of transformative research set against the zeitgeist of the day that the potentials for truly paradigm changing/disruptive scientific exploration are being stifled by the dominance of median-voting, mainstream behaviors held to characterize existing proposal selection processes.

The former perspective is illustrated by the recent National Academies— National Research Council Report: "No theory of scientific progress exists, or is on the horizon, that allows prediction of the future development of new scientific idea or specifies how the different types of scientific progress influence each other-although they clearly are interdependent ...."[54] The latter perspective is embodied in the steady extension, to borrow from van Leeuwen,[55] from descriptive to evaluative to predictive bibliometrics, wherein it is increasingly argued that it is possible to combine big data techniques and citation measures to identify emerging areas of science.[56]

The second contested arena created by the above-described contemporary imperatives for performance-/evidence-based decision-making is likely to take place where the ever-present failures of the scientific community to recognize transformative findings when they occur are opposed to the pressures upon agencies (and performers) to document that investments in transformative research have yielded returns or benefits of whatever sort commensurate with the organizational, programmatic transformations they have caused. Again, in the current and near-term funding environments outlined above, discordant pressures for near(er) term demonstrations of "impacts" relative to the historic rates of acceptance of breakthrough findings are foreseeable, especially if programs to foster transformative research are held to have come at the expense of reduced resources for more traditional (discipline-based) lines of inquiry.

The rate at which and the reasons why new transformative findings are accepted, resisted, or rejected by relevant scientific communities is a topic unto itself. One need not totally ascribe to or reject Planck's well-known aphorism about new ideas triumphing only when adherents of earlier propositions have died off to acknowledge striking examples of delayed recognition of theoretical or empirical findings that have subsequently transformed a field as with the theory of oncogenes.[57] The reception accorded game theory, a now pervasive analytical approach in the social sciences and other fields, and for which John Nash received the Nobel Prize in Economics, is a classic, and indeed telling example, of premature rejection. As noted by Luce and Raiffa in their 1957 survey of the field of decision- making: "We have the historical fact that many social scientists have become disillusioned with game theory. Initially there was a naïve bandwagon feeling that game theory solved innumerable problems of sociology and economics, or that, at least it made their solution a practical matter of a few years' work. This has not turned out to be the case."[58]

Essentially, the above is but an outline of the policy gauntlet that initiatives to promote potentially transformative research will need to course, one made longer and narrower by financial austerity. By itself, however the analysis says little about the merits of the case for these initiatives as a means to foster scientific renewal or the form that any such initiatives might or should take. The policy setting is seen here as too recent and fluid to permit other than programmatic advocacy and analytical speculation.

A more fruitful if oblique approach to assessing the policy dynamism in proposals on behalf of transformative research is to pose a different question: why now? The question is of research interest in its own right. This question connects to a long-standing, fruitful line of inquiry in political science about agenda setting.[59] It has direct bearing here as it relates to the depth and breadth of the changes necessary, or deemed so, to promote the renewal of science to be brought about by initiatives to foster transformative research. For it is one thing to see the pursuit of transformative research as a new permanent policy imperative requiring fundamental structural changes in the priorities and ways in which national research systems are organized, say in changing allocations among fields of science, performers, and funding mechanisms, all those issues at the core of the search for a new science of science policy,[60] and what may very well prove to be a short-lived policy epicycle readily accommodated at the margins of science policy by addressing the immediate and most publicized needs of those advocating for adoption of new approaches.

Policy waves associated with new or fluctuating national priorities are recurring events in the evolution of a nation's science policies and organizational structures.[61] Taking the language employed in the 1945 "The Endless Frontier" as a starting point, the rhetorical landscape of US science policy has changed frequently over the past 70-plus years. The basic research/applied research/development typology is enshrined in the National Science Board's biennial Science and Engineering Indicators, but this classification trinity mainly reflects compliance with the OECD's Frascati Manual, which serves internationally to standardize the formats in which governments report data. Floating about this typology is an ever-changing number of conceptual models and phrases. A short, ready-at-hand listing of adjectives and nouns to describe the objectives and contents of national science policy research would include at least the following: pure, strategic, strategic basic, fundamental, mission (non-mission) oriented, Bohr/Pasteur Quadrants,[62] Translational, Basic Technological Research,[63] Need Driven/Curiosity Driven, Mode 1/Mode 2.

These changes reflect in part the continuous desire and need on the part of sponsors and performers to accurately describe the interrelationships among the constituent activities of the research and development enterprise, especially needed at times to purge discourse or rid mental maps of the tyranny of linear models of innovation. In part too they reflect the claims of performers, such as the engineering community, who have perceived themselves disadvantaged (in funding or status) by the Bush report's heavy emphasis on basic or pure research.

In the main though, the most powerful propellant to this ever-changing vocabulary is the ever-present need of the scientific community to articulate a contemporaneously relevant and persuasive rationale for why public sector funds should be used to support its activities. This dynamic, for example, account for the surge and then fall off in the usage of the metaphor of Pasteur's Quadrant with its evocative claim that basic/pure/or fundamental research (Bohr's Quadrant) is not science for science's sake but is inherently linked to, and often is the prime mover, in the development of new knowledge of use/relevance to society. It accounts too for the rise of the phrase and programmatic initiatives directed at fostering translational research. Filtered through this historical perspective, transformative research appears yet but another policy epicycle.

But why now? The answer suggested here lies in the love–hate relationship felt toward peer or merit review by funding agencies and US academic scientists. Merit review as practiced by the NSF is extolled as "…an international gold standard for review of science and engineering research proposals."[64] Likewise, America's life scientists have credited

NIH's single investigator-initiated, peer review system model of research grant allocation and funding (in contrast to the bloc grant, institutionally based, governmentally determined targeted research foci of many other OECD nations) as contributing to US scientific leadership.[65]

A cottage industry exists on the defects of peer review procedures. Among the most frequent and salient of the criticisms made of the mechanism are the manifest or latent biases it exhibits toward specific populations of researchers (gender, race, age, geographical, or institutional location).[66] Of more immediate relevance here are claims that the mechanism tends to lead to medium-voter behavior, driving out not only poor proposals but also those that depart significantly from what are held to mainstream theories and methods of the set of experts who happen to be convened to review a particular set of proposals.

Assertions that peer review procedures are (unduly) conservative, squeezing out high-risk research clearly are not new. In 1992, James McCullough, staff director of program evaluation at NSF, reported on a 1986 survey of principal investigators (PIs) to which over 9000 PIs responded. According to McCullough two-thirds of the respondents (about 6000) "agreed with the statement that NSF is not likely to fund high-risk, exploratory research because the likelihood of obtaining favorable reviews is slim."[67] Thus, allowing for some modest program adjustments, such as its Small Grants for Exploratory Research program, that empowered program managers to make discretionary awards for proposals that included preliminary work on untested and novel ideas without proposal review, a 20-year interregnum exists between the date of the NSF PI survey and the 2007 publication of "Enhancing Transformative Research." No evidence exists to indicate that the "conservatism" of the peer review process has increased (worsened), or that more breakthrough ideas have been rejected by review panels and/ or program managers or remain stillborn at the pre-proposal stage because researchers believed that this would not be funded.

At their core, the statements about the stifling of transformative research represent expressions of discontent with peer review procedures and existing funding arrangements.

As stated clearly in the NSB 2007 report, "Transformative research frequently does not fit comfortably within the scope of project-focused, innovative, step-by-step research or even major centers, nor does it tend to fare well wherever a review system is dominated by experts highly invested in current paradigms or during times of especially limited budgets that promote aversion to risk."[68]

Expressed thusly, promotion of transformative research represents an attempt to change the terms on which research support is awarded more than an empirical demonstration of retardation in the rate of scientific discovery, however measured. Strikingly, and in a sense counterintuitive to studies of scientific advance that emphasize the singular contributions of relatively younger researchers,[69] the emergence of transformative research as a science policy agenda issue in the USA largely mirrors expressions of discontent by senior scientists to NIH, NSF, and Congress about their inability to get funding for their frontier/interdisciplinary research because of the (perceived/asserted) conservatism of disciplinary/mainstream peer review processes. Having "been there, done that" within the "mainstreams" of their disciplines/paradigms, it is these researchers, seeking to advance, or breakthrough, the boundaries of disciplines they know all too well, who have found their proposed new work rejected by a NSF's gold standard merit review system.

Although the chapter has focused on the USA, it is informative to note that a similar dynamic also appears to hold in the countries included in Heinze's survey of funding schemes for sponsoring ground-breaking research.[70] He writes: "Current funding mechanisms, it seems, are not flexible enough to accept that scientists with excellent track records in their existing fields are capable of investigating phenomena that involve moving into new fields and that there are synergies in funding such research."[71]

If the underlying root cause for the stifling of transformative research lies in the dynamics of peer review processes, how transformative must the changes be in these processes to foster the renewal of science? Subsumed in this question is yet another one: Are such changes, whatever their form, only necessary, or are they also sufficient? A considerable number of alternative proposal selection arrangements immediately present themselves. Included here as a ready if standard set of options are special programmatic initiatives to promote "transformative" research; reconfigured, more interdisciplinary review panels, populated by researchers who have demonstrated success with transformative research, however that is defined by the sponsor; explicit instructions to review panels to pinpoint those proposals which they consider to have breakthrough potential; and adding "potentially transformative" in program solicitations as a selection criterion. Early evidence suggests that each of these options is already in place.

Two additional options, more far reaching in their deviation from existing mainstream techniques, also are readily identifiable. These are (1) allocating funds to individuals based on their established record for breakthrough research rather than on the basis of specific proposals, and (2) increased willingness on the part of program managers to exercise authority they may currently possess but infrequently use to make recommendations and push for proposals they view as potentially transformative even if these recommendations differ from those of expert review panels. In effect, these latter two options involve having science-oriented government agencies, such as NIH and NSF, adopt procedures long identified with DARPA as well as those of foundations. NIH, in fact, has begun to move in that direction: people rather than projects, offering, as in its Pioneer Award program, longer term, increased levels of funding to a competitively selected number of researchers based on their track record and future promise. Consideration for expanded use of this funding model is underway across NIH Institutes, even at the expense of having awards allocated on this basis reduce that available to support individual projects.[72]

Simple in design as these last two options may appear, they fly in the face of other concerns about the "graying" of the scientific workforce, as well as the potential for the further concentration of Federal academic research funding among institutions and states, a perennial source of background contention given the distributive politics characteristic of democratic governments from which agencies derive their funding. For this reason, whatever may be their attractiveness and effectiveness as a means of fostering transformative research, the likelihood that either of these two options will substantially displace existing procedures is low.

9.6 CONCLUSION

Interdependence between funders and performers conditions the conclusions that can be drawn from the above analysis. Starting this time with the behavior of universities, statements about the interdisciplinary course of scientific inquiry (and graduate education) are rife with high degrees of uncertainty. As cited above, many knowledgeable observers continue to advise that staying in one's long cultivated disciplinary garden is the best way to produce the fruits of scientific discovery. Moreover, even to the extent that the future pathways to scientific discovery require increased integration, borrowing, and collaboration across disciplinary boundaries, there yet remain unanswered questions about the optimal organizational arrangements for conducting and evaluating transformative research.

Given the competitive, prestige driven character of the American research university system,[73] answers to these questions seem likely to evolve less from hortatory pronouncements or the activities of any single university in implementing a strategic plan but rather through a Darwinian process involving the interactions among a large number of institutions. If interdisciplinarity in fact is an essential requirement for significant scientific advance, those universities that most rapidly and substantially adjust their institutional priorities and organizational arrangements to accommodate it will be those that gain (or retain) reputation, prominence, and resources (however measured), while those that adapt more slowly or less well will fail to gain or lose position.[74]

If, however, interdisciplinarity proves not to produce the high impacts for scientific renewal or societal relevance projected for it, its institutional impacts are likely be quite modest, readily internalized into university operations, as they have been now for several decades, by establishing various forms of centers and institutes, but not affecting existing college and department arrangements, promotion and tenure criteria, and the like. Once more, the permanence and essentiality of discipline-based departments will have been validated. In either case, universities essentially are making bets on the future. Those who bet heavily on the first outcome will gain in absolute and relative terms compared with those who adapt more slowly; those who adopt the second strategy gain if it emerges as the dominant form, while the others find themselves with relatively little for their initiatives.

But the story does not end here. Instead, it wraps around questions of where the incentives and funding opportunities to conduct interdisciplinary research will come from. The answer to these questions in turn, revolves about future commitments of foundations, federal government agencies, and firms. Given the historic and continuing importance of these external sponsors in providing the discretionary resources and scientific legitimacy often indispensable to the establishment of interdisciplinary research and graduate degree programs, the scale and sustainability of current university thrusts will depend heavily on the outcomes of the internal debates that these organizations are currently engaged in about programmatic priorities between disciplinary and interdisciplinary modes of knowledge creation and transmission. Based on historic experience, without federal or foundation funding, new fields are like small, start-up businesses that invariably enter valleys of death in which they perish unless receiving additional, external funds. Given projected decelerated rates of funding for the major non-defense federal science agencies, intensified competition between single investigator/single disciple and programmatic/interdisciplinary modes of funding, is likely. Thus, the future status of interdisciplinarity on US universities seems as much dependent on priorities, negotiations, and resource allocation decisions in multiple corridors of power as it does on the strategies and activities within the halls of ivy.

Comparable projections on the accommodations that may be required to foster transformative research is more conjectural since the concept remains gumby-like, taking on many shapes. Additionally, agencies have yet to sort through how to integrate, or splice, transformative research initiatives into, atop, or alongside existing programmatic structures. Perhaps the safest projection is that given the protean quality of academic institutions, whatever transformative may be and however long or short its day in the sun as a science policy imperative and source of new funding, in the short term at least it is not likely to transform the way universities operate.

NOTES

1. Benjamin F. Jones, '*As Science Evolves, How Can Science Policy?*' *National Bureau of Economic Research Working Paper 16002* (Cambridge, MA: National Bureau of Economic Research, 2010).

2. Daryl E. Chubin et al., *AAAS Review of the NSF Science and Technology Centers Integrative Partnership (STC) Program, 2000–2009, Report to the National Science Foundation under Grant No. 0949599* (Washington, DC: American Association for the Advancement of Science, 2010).

3. Robert P. Crease, "Physical Sciences," in *The Oxford Handbook of Interdisciplinarity*, ed. R. Frodeman et al. (Oxford: Oxford University Press, 2010); Warren Burggren et al., "Biological Sciences," in *The Oxford Handbook of Interdisciplinarity*, ed. R. Frodeman et al. (Oxford: Oxford University Press, 2010).

4. U.S. Congress, Office of Technology Assessment, *Federally Funded Research: Decisions for a Decade* (Washington, DC: U.S. Government Printing Office, 1991).

5. Susan L. Cutter, "The Vulnerability of Science and the Science of Vulnerability," *Annals of the Association of American Geographers* 93 (2003).

6. Quoted in Philippe Laredo, and Philippe Mustar, ed. *Research and Innovation in the New Global Economy* (Cheltenham, UK: Edward Elgar, 2001).

7. Gunnar Öquist and Mats Benner, *Fostering Breakthrough Research: A Comparative Study* (Stockholm, Sweden: Akademirapport, Kungl. Vetenskapsakademien, 2012).

8. Irwin Feller, "Performance Measurement and the Governance of American Academic Science," *Minerva* 47 (2009).

9. Robert E. Kohler, *Partners in Science* (Chicago, IL: University of Chicago Press, 1991).
10. Julie Klein, *Interdisciplinarity: History, Theory, and Practice* (Detroit, MI: Wayne State University Press, 1990).
11. Julius J. Harwood, "Emergence of the field and early hopes," in *Materials science and engineering in the United States*, ed. R. Roy (University Park, PA: Pennsylvania State University Press, 1969).
12. Ibid.
13. Clark A. Miller, "Policy Challenges and University Reform," in *The Oxford Handbook of Interdisciplinarity*, ed. R. Frodeman et al. (Oxford: Oxford University Press, 2010), 337.
14. Harvey M. Sapolsky and Mark Z. Taylor, "Politics and the Science of Science Policy," in *The Science of Science Policy*, ed. K. Fealing et al. (Stanford, CA: Stanford University Press, 2011); David Goldston, "Science Policy and the Congress," in *The Science of Science Policy*, eds. K. Fealing et al. (Stanford, CA: Stanford University Press, 2011).
15. Irwin Feller, "Interdisciplinarity: Paths Taken and Not Taken," *Change* (2007a).
16. Steven Weinberg, "The Crisis of Big Science," *New York Review of Books*, May 10, 2012.
17. Peter Weingart, "New Modes of Knowledge Production," in *The Oxford Handbook of Interdisciplinarity*, ed. R. Frodeman et al. (Oxford: Oxford University Press, 2010).
18. Jerry Jacobs, "Interdisciplinary Hype," *Chronicle Review*, November 27 (2009): B4–5.
19. Benjamin Mueller, "Hiring that Crosses Disciplines can Create Tensions," *Chronicle of Higher Education*, February 24, 2014.
20. Denise Caruso and Diane Rothen, *Lead, follow, get out of the way: Sidestepping the barriers to effective practice of interdisciplinarity* (San Francisco, CA: Hybrid Vigor Institute, 2001).
21. Barbara Casassus, "Put Focus Back on Basic Research, Say Science Unions," *Nature/News*, accessed October 7, 2014, http://www.nature.com/news/put-focus-back-on-basic-research-say-science-unions-1.15817.
22. Myra H. Strober, *Interdisciplinary Conversations* (Stanford, CA: Stanford University Press, 2011).
23. See Heinze and Jappe, in this volume.
24. Grit Laudel and Gloria Origgi, "Special Issue on the Assessment of Interdisciplinary Research Assessment," *Research Evaluation* 15 (2006); Irwin Feller, "Multiple Actors, Multiple Settings, Multiple Criteria: Issues in Assessing Interdisciplinary Research," *Research Evaluation* 15 (2006); Veronica B. Boix-Mansilla, "Assessing Expert Interdisciplinary Work at the Frontier: An Empirical Exploration," *Research Evaluation* 15 (2006);

Katri Huutoniemi, "Evaluating Interdisciplinary Research," in *The Oxford Handbook of Interdisciplinarity*, ed. R. Frodeman et al. (Oxford: Oxford University Press, 2010).

25. Richard Whitley, "Changing Governance and Authority Relations in the Public Sciences," *Minerva* 49 (2011): 375.

26. Alan L. Porter and Ismael Rafols, "Is Science Becoming More Interdisciplinary? Measuring and Mapping Six Research Fields Over Time," *Scientometrics* 81 (2009); Caroline J. Wagner et al., "Approaches to Understanding and Measuring Interdisciplinary Scientific Research (IDR): A Review of the Literature," *Journal of Informetrics* 165 (2011).

27. Michael Gibbons et al., *The new production of knowledge* (London: SAGE Publications, 1994), 14.

28. Rita R. Colwell, *The National Science Foundation's role in the Artic. Opportunities in Artic research: A community workshop. September 3* (Washington, DC: National Science Foundation, 1998); Jeffrey Brainard, "U.S. agencies look to interdisciplinary science," *Chronicle of Higher Education* June 14, 2002.

29. Norman Metzger and Richard N. Zare, "Science Policy: Interdisciplinary Research: From Belief to Reality," *Science* 283 (1999).

30. Irwin Feller, "New organizations, old cultures: Strategy and implementation of interdisciplinary programs," *Research Evaluation* 11 (2002).

31. George Keller, *Academic strategic planning: The management revolution in American higher education* (Baltimore, MD: Johns Hopkins University Press, 1983); Robert Birnbaum, *Management Fads in Higher Education* (San Francisco, CA: Jossey-Bass, 2000).

32. The 2010 NRC, Assessment of Research Doctorate Programs, based on data from 2006, covered 62 fields, which allowed for increased coverage of newly emerging fields of study. However problems associated with delays in the report's issuance and its substitution of a complex mix of rating systems in lieu of a single number and ranking appear though to have reduced its metric employed in the shaping of an institutions' strategic thinking.

33. Government-University-Industry Research Roundtable, *Stress on research and education at colleges and universities: Institutional and sponsoring agency responses* (Washington, DC: National Academies Press, 1994).

34. Duke University, *Building on excellence: The university plan* (Durham, NC, 2001), 16.

35. Ibid.

36. University of Wisconsin-Madison, *For Wisconsin and the World; Strategic Framework 2009–2014* (Board of Regents of the University of Wisconsin System, 2009), 4.

37. Rhoades, Gary, "Who's Doing it Right? Strategic Activity in Public Research Universities," *Review of Higher Education* 24 (2002); Feller, "Interdisciplinarity."

38. Peter Weingart, "Interdisciplinarity: The paradoxical discourse," in *Practising interdisciplinarity*, ed. Peter Weingart and Nico Stehr (Toronto, Canada: University of Toronto Press, 2000); Stephen Turner, "What are disciplines? And how is interdisciplinarity different?," in *Practising interdisciplinarity*, ed. Peter Weingart and Nico Stehr (Toronto, Canada: University of Toronto Press, 2000); Andrew Abbott, "The Disciplines and the Future," in *The Future of the City of Intellect*, ed. Steven G. Brint (Palo Alto, CA: Stanford University Press, 2002).

39. Abbott, "The Disciplines," 210.

40. Ibid., 218.

41. Quoted in Morton Keller and Phyllis Keller, *Making Harvard Modern* (Cambridge, MA: Harvard University Press, 2001), 6.

42. John Servos, *Physical chemistry from Ostwald to Pauling* (Princeton, NJ: Princeton University Press, 1990), xiv.

43. George J. Stigler, "Specialism: A dissenting opinion," in *The intellectual and the marketplace, and other essays*, ed. George J. Stigler (New York: The Free Press of Glencoe, 1963), 11.

44. Ashley Mooney, "Dismantling the Duke Institute for Genome Sciences & Policy," Online at http://www.dukechronicle.com/articles/2014/04/10/dismantling-duke-institute-genome-sciences-policy#.Vcn8p_led68.

45. Christina Raasch et al., "The Rise and Fall of Interdisciplinary Research: The Case of Open Source Innovation," *Research Policy* 42 (2013).

46. Diana Rhoten, *A multi-method analysis of social and technical conditions for interdisciplinary collaboration. Final Report to the National Science Foundation, Grant #BCS-0129573* (San Francisco, CA: The Hybrid Vigor Institute, 2003), 42.

47. Weingart, "New Modes," 13.

48. James J. Duderstadt, *A University for the 21st Century* (Ann Arbor, MI: University of Michigan Press, 2000), 120.

49. Theodore R. Mitchell, "Border crossings: Organizational boundaries and challenges to the American professoriate," *Daedalus* (1997).

50. David L. Kirp, *Shakespeare, Einstein, and the bottom line* (Cambridge, MA: Harvard University Press, 2003), Chapter 4.

51. Keller and Keller, *Making Harvard Modern*, 498.

52. B. Lal, *Transformative Research—An Exploration of Six Propositions* (Unpublished, 2012), 9.

53. James S. Dietz and Juan. D Rogers, "Meanings and Policy Implications of 'Transformative Research': Frontiers, Hot Science, Evolution, and Investment Risk," *Minerva* 50 (2012): 22.

54. National Academies—National Research Council, *A Strategy for Assessing Science* (Washington, DC: National Academies Press, 2007), 73.

55. Thed Van Leeuwen, "Descriptive versus Evaluative Bibliometrics," in *Handbook of Quantitative Science and Technology Research*, ed. H. Moed, et al. (Dordrecht: Kluwer Academic Publishers, 2004).

56. Evans, while generally positive about the potential for such models to accelerate the transitions from time of publication to widespread acceptance, has also cautioned however about the risks of self-filling prophecies generated by acceptance of their findings. His concern is that "The swarming of researchers and sponsors about a predicted hot topic could lead them to 'prematurely abort ideas that have a second or third act to play.'" James A. Evans, "Future Science," *Science* 342 (2013): 44. Also, Henry Small et al., "Identifying Emerging Topics in Science and Technology," *Research Policy* 43 (2014).

57. Daniel Kevles, "Pursuing the Unpopular: A History of Viruses, Courage and Cancer," in *Hidden Histories of Science*, ed. Robert Silvers (New York: A New York Review Book, 1995).

58. Quoted in Sylvia Nasar, *A Beautiful Mind* (New York: Simon and Schuster, 1998), 122.

59. Jack L. Walker, "The Diffusion of Innovations Among the American States," *American Journal of Political Science Review* 63 (1977); Frank R. Baumgartner and Bryan D. Jones, *Agendas and Instability in American Politics* (Chicago: University of Chicago Press, 1993); John W. Kingdon, *Agendas, Alternatives, and Public Policies* (New York: Harpers Collins College Publishers, 1995).

60. John H. Marburger, "Why Policy Implementation Needs a Science of Science Policy," in *The Science of Science Policy*, ed. K. Fealing et al., (Stanford, CA: Stanford University Press, 2011).

61. William B. Bonvillian, "The New Model Innovation Agencies: An Overview," *Science and Public Policy* 41 (2014).

62. Donald Stokes, *Pasteur's Quadrant* (Washington, DC: Brookings Institution, 1997).

63. Lewis M. Branscomb, "From Technology Politics to Technology Policy," *Issues in Science and Technology* 13 (1997).

64. National Science Board, *Enhancing Support for Transformative Research at the National Science Foundation* (Arlington, VA: National Science Foundation, 2007), 4.

65. National Academies, *Experiments in International Benchmarking of US Research Fields* (Washington, DC: National Academies Press, 2000), 3–20.

66. Daryl E. Chubin and Edward J. Hackett, *Peerless Science* (Albany, NY: State University of New York Press, 1990); Susan E. Cozzens,"Death by Peer Review," in *The Changing Governance of Science*, ed. Richard Whitley

and Jochen Gläser (Dordrecht: Springer, 2007); Irwin Feller, "Peer Review and Expert Panels as Techniques for Evaluating the Quality of Academic Research," in *Handbook on the Theory and Practice of Program Evaluation*, ed. A. Link and N. Vonortas (Cheltenham, UK: Edward Elgar Publishers,, 2013); Stephen A. Gallo et al., "The Validation of Peer Review Through Research Impact: Implications for Funding Strategies," *PLoS One* 9, no. 9 (2014).

67. J. McCullough, "Federal Peer Review Systems: Too Conservative to Support High-Risk Research," in *AAAS Science and Technology Policy Yearbook—1992*, ed. S. Nelson et al. (Washington, DC: American Association for the Advancement of Science, 1993), 246.

68. National Science Board, *Enhancing Support*, 4.

69. Paula F. Stephan and Sharon G. Levin, *Striking the Mother Lode in Science* (New York: Oxford University Press, 1992).

70. Thomas Heinze, "How to Sponsor Ground-breaking Research: A Comparison of Funding Schemes," *Science and Public Policy* 35 (2008).

71. Ibid., 305.

72. Jocelyn Kaiser, "NIH Institute Considers Broad Shift to 'People' Awards," *Science* 345 (2014).

73. Henry Rosovsky, *The university: An owner's manual* (Cambridge, MA: Harvard University Press, 1990); Clark Kerr, "The New Race to be Harvard or Berkeley or Stanford," *Change* 23 (1993).

74. Irwin Feller, "Who Races with Whom? Who is Likely to Win (or Survive)? Why?," in *Future of the American Research Economy*, ed. R. Geiger et al. (Rotterdam: Sense Publishers, 2007b); Heinze and Jappe, in this volume.

Acknowledgments Research on this chapter was supported by grants from Atlantic Philanthropies and by the NSF's Science of Science and Innovation Policy Program (Grant #1441455). I am indebted to my colleagues Daryl Chubin and Pallavi Phartiyal for comments on an earlier version. The views expressed in this chapter are mine alone.

REFERENCES

Abbott, Andrew. 2002. The disciplines and the future. In *The future of the city of intellect*, ed. Steven G. Brint, 205–230. Palo Alto: Stanford University Press.

Baumgartner, Frank R., and Bryan D. Jones. 1993. *Agendas and instability in American politics*. Chicago: University of Chicago Press.

Birnbaum, Robert. 2000. *Management fads in higher education*. San Francisco: Jossey-Bass.

Boix-Mansilla, Veronica B. 2006. Assessing expert interdisciplinary work at the frontier: An empirical exploration. *Research Evaluation* 15: 17–29.

Bonvillian, William B. 2014. The new model innovation agencies: An overview. *Science and Public Policy* 41: 425–437.

Brainard, Jeffrey. 2002. U.S. agencies look to interdisciplinary science. *Chronicle of Higher Education*, June 14, p. A20ff.

Branscomb, Lewis M. 1997. From technology politics to technology policy. *Issues in Science and Technology* 13: 41–48.

Burggren, Warren, Kent Chapman, Bradley Keller, Michael Monticino, and J. Torday. 2010. Biological sciences. In *The Oxford handbook of interdisciplinarity*, ed. R. Frodeman, J. Klein, C. Mitcham, and J. Holbrook, 119–131. Oxford: Oxford University Press.

Caruso, Denise, and Diane Rothen. 2001. *Lead, follow, get out of the way: Sidestepping the barriers to effective practice of interdisciplinarity*. San Francisco: Hybrid Vigor Institute.

Casassus, Barbara. Put focus back on basic research, say science unions. *Nature/ News*. http://www.nature.com/news/put-focus-back-on-basic-research-say-science-unions-1.15817. Accessed 7 Oct 2014.

Chubin, Daryl E., and Edward J. Hackett. 1990. *Peerless science*. Albany: State University of New York Press.

Chubin, Daryl E., Edward Derrick, Irwin Feller, and Pallavi Phartiyal. 2010. *AAAS review of the NSF Science and Technology Centers Integrative Partnership (STC) program, 2000–2009, Report to the National Science Foundation under Grant No. 0949599*. Washington, DC: American Association for the Advancement of Science.

Colwell, Rita R. 1998. *The National Science Foundation's role in the Artic. Opportunities in Artic research: A community workshop. September 3*. Washington, DC: National Science Foundation.

Cozzens, Susan E. 2007. Death by peer review. In *The changing governance of science*, ed. Richard Whitley and Jochen Gläser, 225–242. Dordrecht: Springer.

Crease, Robert P. 2010. Physical sciences. In *The Oxford handbook of interdisciplinarity*, ed. R. Frodeman, J. Klein, C. Mitcham, and J. Holbrook, 79–102. Oxford: Oxford University Press.

Cutter, Susan L. 2003. The vulnerability of science and the science of vulnerability. *Annals of the Association of American Geographers* 93: 1–12.

Dietz, James S., and Juan D. Rogers. 2012. Meanings and policy implications of 'transformative research': Frontiers, hot science, evolution, and investment risk. *Minerva* 50: 21–44.

Duderstadt, James J. 2000. *A university for the 21st century*. Ann Arbor: University of Michigan Press.

Duke University. 2001. *Building on excellence: The university plan*. Durham: Duke University.

Evans, James A. 2013. Future science. *Science* 342: 44–45.

Feller, Irwin. 2002. New organizations, old cultures: Strategy and implementation of interdisciplinary programs. *Research Evaluation* 11: 109–116.

Feller, Irwin. 2006. Multiple actors, multiple settings, multiple criteria: Issues in assessing interdisciplinary research. *Research Evaluation* 15: 5–15.

Feller, Irwin. 2007a. Interdisciplinarity: Paths taken and not taken. *Change* 39: 46–51.

Feller, Irwin. 2007b. Who races with whom? Who is likely to win (or survive)? Why? In *Future of the American research economy*, ed. R. Geiger, C. Colbeck, R. Williams, and C. Anderson, 71–90. Rotterdam: Sense Publishers.

Feller, Irwin. 2009. Performance measurement and the governance of American academic science. *Minerva* 47: 323–344.

Feller, Irwin. 2013. Peer review and expert panels as techniques for evaluating the quality of academic research. In *Handbook on the theory and practice of program evaluation*, ed. A. Link and N. Vonortas, 115–142. Cheltenham: Edward Elgar Publishers.

Gallo, Stephen A., Afton S. Carpenter, David Irwin, Caitlin D. McPartland, Joseph Travis, Sofie Reynders, Lisa A. Thompson, and Scott R. Glisson. 2014. The validation of peer review through research impact: Implications for funding strategies. *PLoS ONE* 9(9), e106474.

Gibbons, Michael, Camille Limoges, Helga Nowotny, Simon Schwartzman, Peter Scott, and Martin Trow. 1994. *The new production of knowledge*. London: SAGE Publications.

Goldston, David. 2011. Science policy and the Congress. In *The science of science policy*, ed. K. Fealing, J. Lane, J. Marburger, and S. Shipp, 327–336. Stanford: Stanford University Press.

Government-University-Industry Research Roundtable. 1994. *Stress on research and education at colleges and universities: Institutional and sponsoring agency responses*. Washington, DC: National Academies Press.

Harwood, Julius J. 1969. Emergence of the field and early hopes. In *Materials science and engineering in the United States*, ed. R. Roy, 1–10. University Park: Pennsylvania State University Press.

Heinze, Thomas. 2008. How to sponsor ground-breaking research: A comparison of funding schemes. *Science and Public Policy* 35: 302–318.

Huutoniemi, Katri. 2010. Evaluating interdisciplinary research. In *The Oxford handbook of interdisciplinarity*, ed. R. Frodeman, J. Klein, C. Mitcham, and J. Holbrook, 309–320. Oxford: Oxford University Press.

Jacobs, Jerry. 2009. Interdisciplinary hype. *Chronicle Review*, November 27, B4–B5.

Jones, Benjamin F. 2010. *As science evolves, how can science policy?* National Bureau of Economic Research working paper, no. 16002. Cambridge, MA: National Bureau of Economic Research.

Kaiser, Jocelyn. 2014. NIH institute considers broad shift to 'people' awards. *Science* 345: 366–367.

Keller, George. 1983. *Academic strategic planning: The management revolution in American higher education.* Baltimore: Johns Hopkins University Press.

Keller, Morton, and Phyllis Keller. 2001. *Making Harvard modern.* Cambridge, MA: Harvard University Press.

Kerr, Clark. 1993. The new race to be Harvard or Berkeley or Stanford. *Change* 23: 8–15.

Kevles, Daniel. 1995. Pursuing the unpopular: A history of viruses, courage and cancer. In *Hidden histories of science*, ed. Robert Silvers. New York: A New York Review Book.

Kingdon, John W. 1995. *Agendas, alternatives, and public policies.* New York: Harpers Collins College Publishers.

Kirp, David L. 2003. *Shakespeare, Einstein, and the bottom line.* Cambridge, MA: Harvard University Press.

Klein, Julie. 1990. *Interdisciplinarity: History, theory, and practice.* Detroit: Wayne State University Press.

Kohler, Robert E. 1991. *Partners in science.* Chicago: University of Chicago Press.

Lal, B. 2012. *Transformative research—An exploration of six propositions* (Unpublished doctoral dissertation). Washington, DC: George Washington University. Abstract available at: http://phdtree.org/pdf/23418156-transformative-research-an-exploration-of-six-propositions/

Laredo, Philippe, and Philippe Mustar (eds.). 2001. *Research and innovation in the new global economy.* Cheltenham: Edward Elgar.

Laudel, Grit, and Gloria Origgi. 2006. Special issue on the assessment of interdisciplinary research assessment. *Research Evaluation* 15: 2–4.

Mansilla, Veronica B., and Howard Gardner. 2004. Assessing Interdisciplinary Work at the Frontier: An empirical exploration of 'symptoms of quality'. http://www.interdisciplines.org/interdisciplinarity/papers/6. Accessed 26 Feb 2004.

Marburger, John H. 2011. Why policy implementation needs a science of science policy. In *The science of science policy*, ed. K. Fealing, J. Lane, J. Marburger, and S. Shipp, 9–22. Stanford: Stanford University Press.

McCullough, J. 1993. Federal peer review systems: Too conservative to support high-risk research. In *AAAS science and technology policy yearbook—1992*, ed. S. Nelson, K. Gramp, and A. Teich, 245–249. Washington, DC: American Association for the Advancement of Science.

Metzger, Norman, and Richard N. Zare. 1999. Science policy: Interdisciplinary research: From belief to reality. *Science* 283: 642–643.

Miller, Clark A. 2010. Policy challenges and university reform. In *The Oxford handbook of interdisciplinarity*, ed. R. Frodeman, J. Klein, C. Mitcham, and J. Holbrook, 333–344. Oxford: Oxford University Press.

Mitchell, Theodore R. 1997. Border crossings: Organizational boundaries and challenges to the American professoriate. *Daedalus* 126: 265–292.

Mooney, Ashley. 2014. Dismantling the Duke Institute for Genome Sciences & Policy. http://www.dukechronicle.com/articles/2014/04/10/dismantling-duke-institute-genome-sciences-policy#.Vcn8p_led68. Accessed 14 Aug 2015.

Mueller, Benjamin. 2014. Hiring that crosses disciplines can create tensions. *Chronicle of Higher Education*, 24 Feb 2014.

Nasar, Sylvia. 1998. *A beautiful mind.* New York: Simon and Schuster.

National Academies. 2000. *Experiments in international benchmarking of US research fields.* Washington, DC: National Academies Press.

National Academies—National Research Council. 2003. *Assessing research doctorate programs—A methodology study,* ed. J. Ostriker and C. Kuh. Washington, DC: National Academies Press.

National Academies—National Research Council. 2007. *A strategy for assessing science.* Washington, DC: National Academies Press.

National Science Board. 2007. *Enhancing support for transformative research at the National Science Foundation.* Arlington: National Science Foundation.

Öquist, Gunnar, and Mats Benner. 2012. *Fostering breakthrough research: A comparative study.* Stockholm: Akademirapport, Kungl. Vetenskapsakademien.

Porter, Alan L., and Ismael Rafols. 2009. Is science becoming more interdisciplinary? Measuring and mapping six research fields over time. *Scientometrics* 81: 719–743.

Raasch, Christina, Viktor Lee, Sebastian Spaeth, and Cornelius Herstatt. 2013. The rise and fall of interdisciplinary research: The case of open source innovation. *Research Policy* 42: 1138–1152.

Rhoades, Gary. 2002. Who's doing it right? Strategic activity in public research universities. *Review of Higher Education* 24: 41–66.

Rhoten, Diana. 2003. *A multi-method analysis of social and technical conditions for interdisciplinary collaboration. Final report to the National Science Foundation, Grant #BCS-0129573.* San Francisco: The Hybrid Vigor Institute.

Rosovsky, Henry. 1990. *The university: An owner's manual.* Cambridge, MA: Harvard University Press.

Sapolsky, Harvey M., and Mark Z. Taylor. 2011. Politics and the science of science policy. In *The science of science policy,* ed. K. Fealing, J. Lane, J. Marburger, and S. Shipp, 31–55. Stanford: Stanford University Press.

Servos, John. 1990. *Physical chemistry from Ostwald to Pauling.* Princeton: Princeton University Press.

Small, Henry, Kevin W. Boyack, and Richard Klavans. 2014. Identifying emerging topics in science and technology. *Research Policy* 43: 1450–1467.

Stephan, Paula F., and Sharon G. Levin. 1992. *Striking the mother lode in science.* New York: Oxford University Press.

Stigler, George J. 1963. Specialism: A dissenting opinion. In *The intellectual and the marketplace, and other essays,* ed. George J. Stigler. New York: The Free Press of Glencoe.

Stokes, Donald. 1997. *Pasteur's quadrant*. Washington, DC: Brookings Institution.

Strober, Myra H. 2011. *Interdisciplinary conversations*. Stanford: Stanford University Press.

Turner, Stephen. 2000. What are disciplines? And how is interdisciplinarity different? In *Practising interdisciplinarity*, ed. Peter Weingart and Nico Stehr, 46–64. Toronto: University of Toronto Press.

University of Wisconsin-Madison. 2009. *For Wisconsin and the World; Strategic Framework 2009–2014*. Copyright: Board of Regents of the University of Wisconsin System.

U.S. Congress, Office of Technology Assessment. 1991. *Federally funded research: Decisions for a decade*. Washington, DC: U.S. Government Printing Office.

Van Leeuwen, Thed. 2004. Descriptive versus evaluative bibliometrics. In *Handbook of quantitative science and technology research*, ed. H. Moed, W. Glanzel, and U. Schmoch, 373–388. Dordrecht: Kluwer.

Wagner, Caroline, J. David Roessner, Kamau Bobb, Julie T. Klein, Kevin W. Boyack, Joann Keyton, Ismael Rafols, and Katy Börner. 2011. Approaches to understanding and measuring interdisciplinary scientific research (IDR): A review of the literature. *Journal of Informetrics* 165: 14–26.

Walker, Jack L. 1977. The diffusion of innovations among the American states. *American Journal of Political Science Review* 63: 880–899.

Weinberg, Steven. 2012. The crisis of big science. *New York Review of Books*, May 10.

Weingart, Peter. 2000. Interdisciplinarity: The paradoxical discourse. In *Practising interdisciplinarity*, ed. Peter Weingart and Nico Stehr, 25–41. Toronto: University of Toronto Press.

Weingart, Peter. 2010. New modes of knowledge production. In *The Oxford handbook of interdisciplinarity*, ed. R. Frodeman, J. Klein, C. Mitcham, and J. Holbrook, 3–14. Oxford: Oxford University Press.

Whitley, Richard. 2011. Changing governance and authority relations in the public sciences. *Minerva* 49: 359–385.

INDEX

Note: Tables are indicated with *t*, figures with *f*.

© The Editor(s) (if applicable) and The Author(s) 2016
T. Heinze, R. Münch (eds.), *Innovation in Science and
Organizational Renewal*, DOI 10.1057/978-1-137-59420-4

275